Signs of Resurrection

Body #1

Caterpillar

Body #2

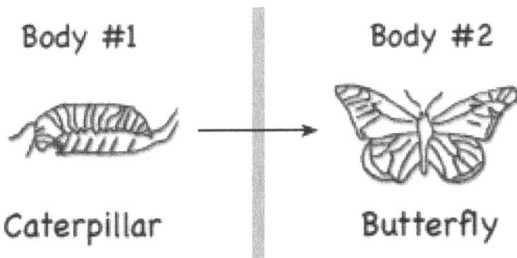

Butterfly

"Just when the caterpillar
thought the world was over,
it became a butterfly."
—Anonymous

Roger Skrenes

En Route Books and Media, LLC
Saint Louis, MO

✹ENROUTE

Make the time

En Route Books and Media, LLC
5705 Rhodes Avenue
St. Louis, MO 63109

Cover credit: Roger Skrenes
Copyright © 2022 Roger Skrenes

ISBN-13: 979-8-88870-015-0
Library of Congress Control Number
2022951734

Resurrection refers to a rising up again to a new life after one has died. It is a new kind of empowered life. The resurrected person who is born into new life cannot die a second time. Resurrection into new life is a kind of stark new beginning.

While others had been raised from the dead into their old lives only to die again, the first and only resurrection event in the history of the world that rose into new life was that of Jesus Christ on the historic day of Easter. The divine Person of Christ could not be held down by death, and his promise to us is that we will not be either.

As the Creator of the world around us, Christ had also put into nature various "signs" of resurrection. This book is written to identify some of those signs or hints of resurrection in nature.

Such a study should help one to see death and our new life beyond death in a new light. This should also move a person to study "the way" of Jesus and his resurrection as recorded in the Bible. Then, by living in this new light, we can hope to fit into God's larger plan – in what we call "Heaven."

Roger Skrenes

Table of Contents

<u>Signs of Risen Life</u>

in…

1 — the universe   1

2 — plants   25

3 — animals   37

4 — human body   63

5 — human behavior   79

6 — love   99

7 — family   111

8 — Christ   131

Final Thoughts   141

Chapter 1

Signs of Risen Life
in the Universe

God's Care

Future	**Heaven**	God creates heaven as a home for his 'spirit-ed' people.
5 M/B.C.	People 5	God creates a 'spirit' for people. People are spirits within a body.
1,800 M/B.C.	Animals 4	God creates animals. Through the history of animals people receive a 'physical' body.
3,500 M/B.C.	Plants 3	God creates plants. Plants make oxygen and food for people.
4,600 M/B.C.	Earth 2	God gives people a place to live while in a physical body.
5,000 Million B.C. (Before Christ)	Sun Stars Stars	God creates the sun giving people light and warmth.

God's Power

Venus   Mercury

Sun

Mars   →Earth

Jupiter

Acts 17,18

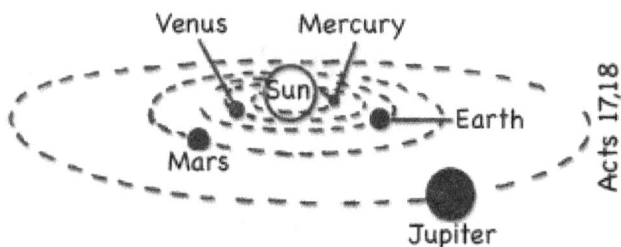

When we see good products in a store we know that some smart person or persons made them. When we look up in the sky we see big products, planets and stars; and we know too that men do not make such things. We also know, deep within us, that some super-person did make planets and stars. This all-powerful maker we call God. And we sense that it is this Big Person who will keep us alive when our physical body wears out. He will do this because, like a Father who brings children into Being, he will protect them and keep them safe (Mt 6,10). Furthermore, since God is eternal so too will be the lives of his children.

Unmoved Mover

All energy comes from God (#1).
This energy (#3-5) makes possible motion
& change in our world (#6-8).

	Me	
People	8	Eat green plants (vegetables) and farm animals (meat) empowering us to move and grow.
Animals	7	Eat green plants as food; that gives them energy to move and grow.
Green plants on Earth	6	Capture sun's energized particles & use them to fix carbon from the air, forming sugar molecules. Long chains of sugar units become starches (food for animals).
Sun	5	Our star which sends energized particles to Earth.
Atoms	4	Are collections of 'part'icles formed inside stars.
Particles	3	Are the 'parts' of atoms that God created from nothing.
Nothing	2	What existed before God created.
God	1	Is the creator, or First Cause. God is uncaused or unmoved by anything else.
	God	

Directions

	Heaven		
0 A.D.	↑ 5 <u>Omniscience</u> appears: All knowing Creator	God— Man, Jesus	heavenly ↑
5 Million B.C. (Before Christ)	4 <u>Intellect</u> appears	Man	intellec- tual ↑
1,800 M/B.C.	3 <u>Senses</u> appear	Animals	psychic ↑
3,500 M/B.C.	2 <u>Life</u> appears	Plants	biological ↑
5,000 M/B.C.	<u>Matter</u> appears ↑ **Earth**	Elements hydrogen, oxygen, etc.	Chemical ↑ Levels within a person

Space Clock

God cares for our space clock every moment.
In like manner, God will care for us in death.

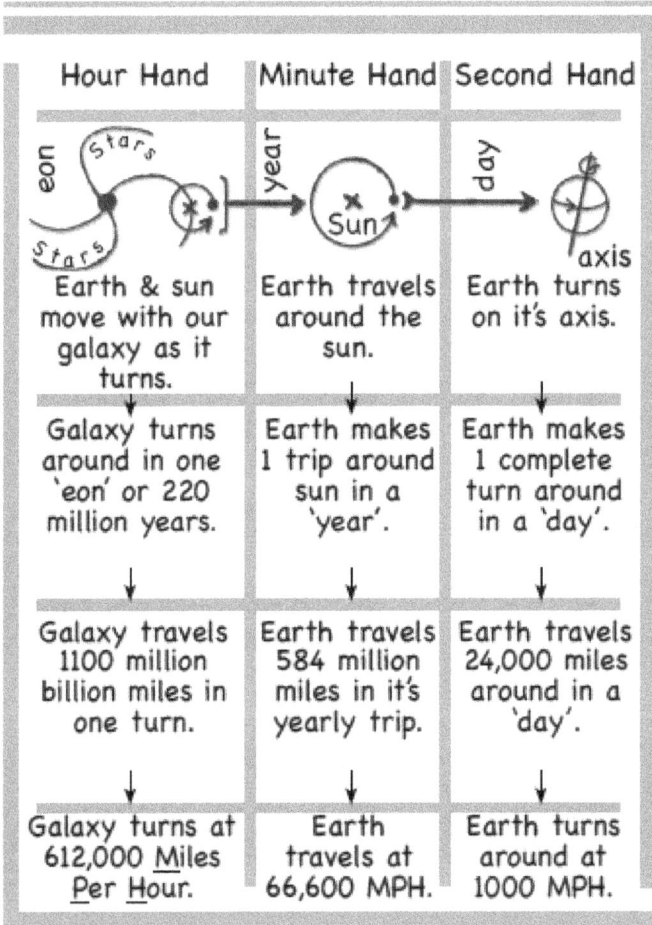

Hour Hand	Minute Hand	Second Hand
Earth & sun move with our galaxy as it turns.	Earth travels around the sun.	Earth turns on it's axis.
Galaxy turns around in one 'eon' or 220 million years.	Earth makes 1 trip around sun in a 'year'.	Earth makes 1 complete turn around in a 'day'.
Galaxy travels 1100 million billion miles in one turn.	Earth travels 584 million miles in it's yearly trip.	Earth travels 24,000 miles around in a 'day'.
Galaxy turns at 612,000 Miles Per Hour.	Earth travels at 66,600 MPH.	Earth turns around at 1000 MPH.

Atoms

Our body is made of atoms.
In each atom there are 3 massless particles: the photon.
neutrino, and gluon. These particles are real but are not a
kind of matter. They act upon particles with mass but have
no mass themselves. Our risen body could be made of this.

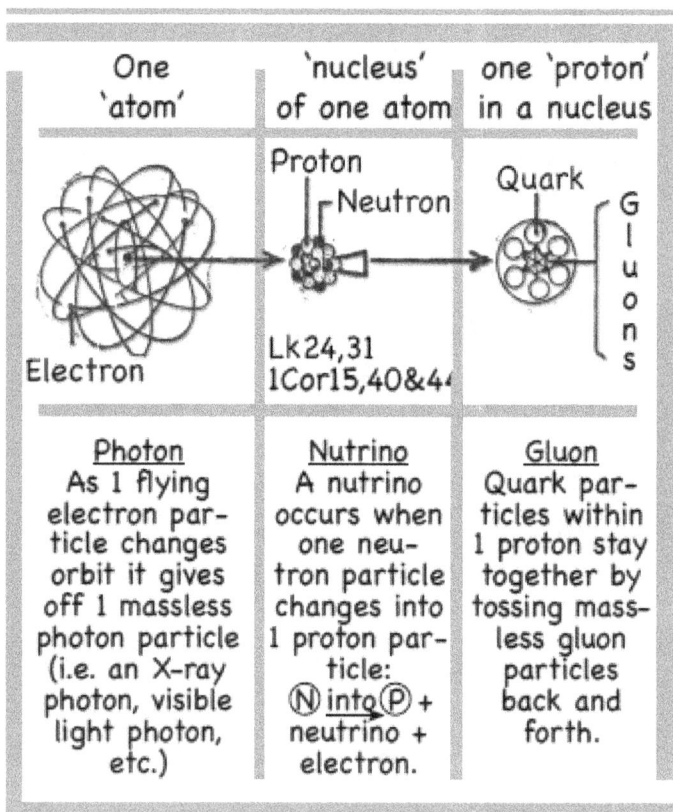

One 'atom'	'nucleus' of one atom	one 'proton' in a nucleus
Electron	Proton ┌Neutron Lk24,31 1Cor15,40&4	Quark G l u o n s
Photon As 1 flying electron particle changes orbit it gives off 1 massless photon particle (i.e. an X-ray photon, visible light photon, etc.)	**Nutrino** A nutrino occurs when one neutron particle changes into 1 proton particle: Ⓝ into Ⓟ + neutrino + electron.	**Gluon** Quark particles within 1 proton stay together by tossing massless gluon particles back and forth.

Radio Waves

In the city

We are constantly hit by radio waves. Radio waves carry the images and voices of people 'broadcast' through the air. Radio and TV people are invisibly present in the air around us. Their presence is real even though we can't see or hear them. To see or hear them we use a radio or TV 'receiver' to 'pick up' their presence.

In like manner we do not see or hear resurrected people. After death, however we will have a new kind of seeing and hearing enabling us to see and speak with other risen people.

Full Moon

1 →	2 →	3 →	4
○	(moon partially behind cloud)	(dashed circle behind cloud)	○
Full moon is seen (No clouds)	Full moon is partially seen (Partially cloud covered)	Full moon is blocked out (Fully cloud covered)	Full moon returns (Clouds are gone)
(standing figure)	(bent figure)	(figure in box)	(risen figure with rays)
Young person	Older person	Person at death	Risen person

Moon Phases

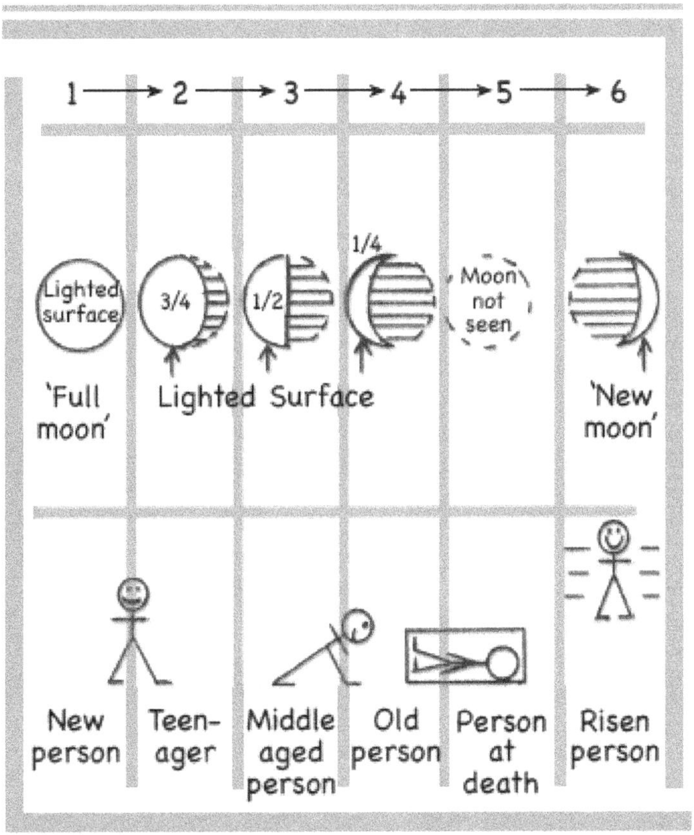

1 → 2 → 3 → 4 → 5 → 6

Lighted surface

3/4

1/2

1/4

Moon not seen

'Full moon'

Lighted Surface

'New moon'

New person

Teen-ager

Middle aged person

Old person

Person at death

Risen person

Moon's Shadow

1 → 2 → 3 → 4			
Ordinary day, full sun	Partial eclipse	Total eclipse	Return of day, full sun
Sun's rays → → → Moon		Earth	
	Partially dark outside during daytime	Totally dark outside during daytime	Resurrection
		Lk 23,44	
Younger person	Older person	Person at death	Risen person

Daylight

3
Winter
north pole
leans away
from sun

2-Fall

Short
day

long
day

1
Summer
north pole
leans
towards sun

4-Spring

Summer ⟶	Fall ⟶	Winter ⟶	Spring
full life	fall in life	death	new life
16 hours daylight. Most of day is light.	12 hours daylight. Darkness begins to set in.	8 hours daylight. Most of day is dark.	12 hours daylight. Light returns.
Young person	Older person	Person at death	Risen Person

Sunrise

Each day the sun goes down (dies)
and then comes up again (rises). This can be
seen as a model of our own death ($\downarrow$)
and resurection ($\uparrow$). God provides this
experience 365 times each year.

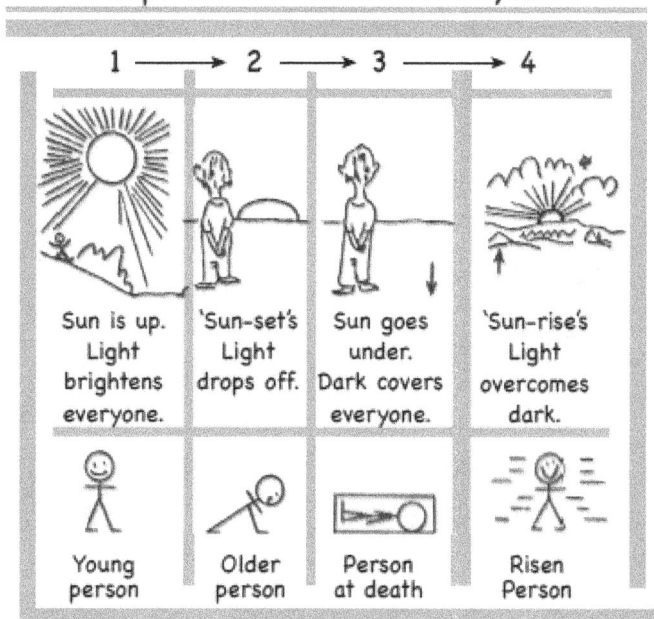

1 →	2 →	3 →	4
Sun is up. Light brightens everyone.	'Sun-set's Light drops off.	Sun goes under. Dark covers everyone.	'Sun-rise's Light overcomes dark.
Young person	Older person	Person at death	Risen Person

Return of Day

"darkness, does the face
of the Earth entomb"
Shakespeare - Macbeth II,4,6

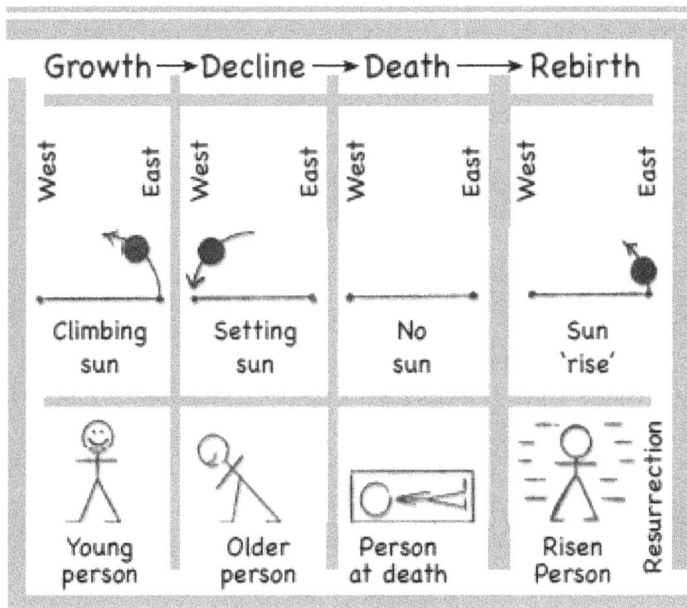

Growth →	Decline →	Death →	Rebirth
West / East	West / East	West / East	West / East
Climbing sun	Setting sun	No sun	Sun 'rise'
Young person	Older person	Person at death	Risen Person — Resurrection

Light

At night the sun seems gone.
When someone dies God seems gone. The sun,
however, is not really gone. The sun merely
shines on the other side of the Earth. Likewise
God is not gone when a person dies. God merely
shines on the other side of that person's life.

Death   Resurrection
(nighttime) ⟶   (daytime)

L.A.   Sun

Earth

Person   Risen
in death   Person

Rain

1 ———→ 2 ———→ 3 ———→ 4

Broken clouds, few in number.	Grey cloud mass, 1 mile thick layer.	Black cloud mass 4 mile thick layer.	No clouds are present.
Broken sunshine	Clouds block sunshine (bright sky)	Clouds block sunlight (dark sky)	Bright sunshine
Young person	Older person	Person at death	Risen Person

Continents Rise

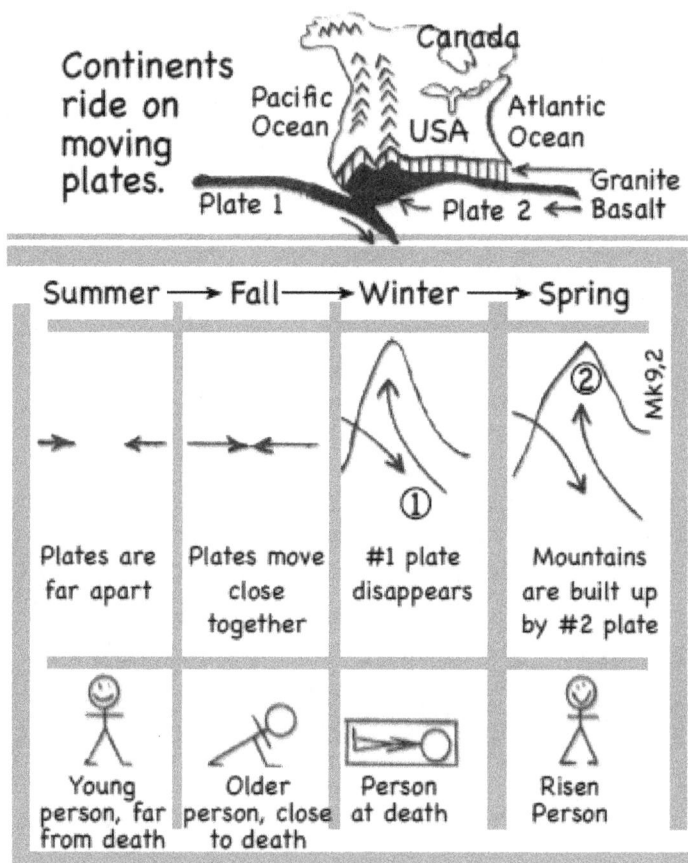

Continents ride on moving plates.

Pacific Ocean

Canada

Atlantic Ocean

USA

Plate 1

Plate 2 ← Basalt

Granite

Summer → Fall → Winter → Spring			
Plates are far apart	Plates move close together	#1 plate disappears	Mountains are built up by #2 plate
Young person, far from death	Older person, close to death	Person at death	Risen Person

① ② Mk9,2

Rocks

Summer →	Fall →	Winter →	Spring
full life	fall in life	death	new life
Sand on surface	Sand particles are buried	Sandstone is formed by heat	Quartzite rock
gets covered up		heat	Sandstone is raised up to form metamorphic quartzite
Young person	Older person	Person at death	Risen Person

Water (H$_2$O)

Summer →	Fall →	Winter →	Spring
Water vapor (gas)	Flowing 'water' (liquid)	Water ice (solid)	Water melting & rising
moving fast	Water molecules slowing down	stilled motion	motion unlocked
Young person	Older person	Person at death	Risen Person

The Sea

Summer →	Fall →	Winter →	Spring
full life	fall in life	death	new life
Normal water	Rough water	Tumultuous water	Calm water
Small waves	Large waves (White caps)	Very large waves	Ripples
Young person	Older person	Person at death	Risen Person

Mt 8,24

Clouds

Steam is water that we cannot see.
Water is changed into steam as the sun heats
it. Steam 'rises up' because it weighs very little.
After rising, it cools and changes back into tiny
water drops. These water drops form clouds.

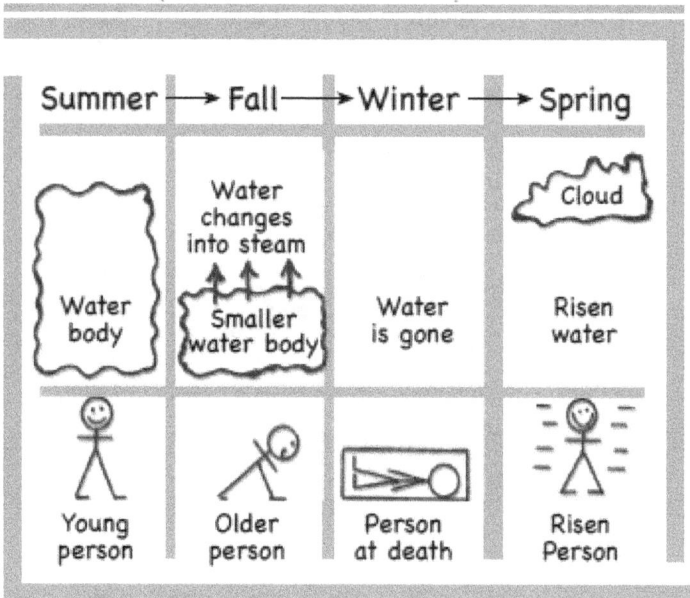

Summer	Fall	Winter	Spring
Water body	Water changes into steam / Smaller water body	Water is gone	Cloud / Risen water
Young person	Older person	Person at death	Risen Person

Wind

Summer →	Fall →	Winter →	Spring
Ordinary gentle breeze	Gusty wind	Stormy wind	Calm (no wind)
5 MPH	20 MPH	40 MPH	1 MPH
young person	Older person	Person at death	Risen Person

Oxygen

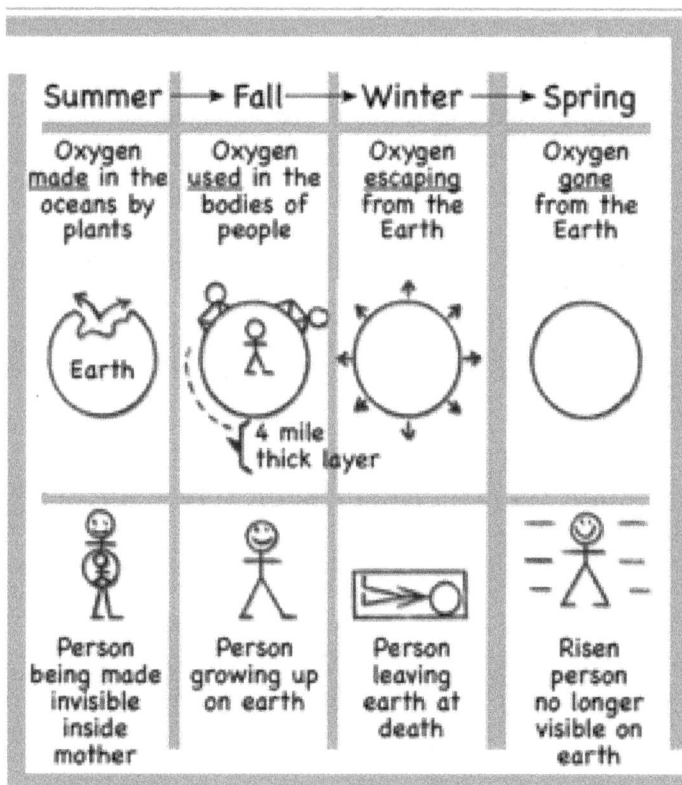

Chapter 2

Signs of Risen Life
in Plants

God's Care

Future	**Heaven**	God creates heaven as a home for his 'spirit-ed' people.
5 M/B.C.	People 5	God creates a 'spirit' for people. People are spirits within a body.
1,800 M/B.C.	Animals 4	God creates animals. Through the history of animals people receive a 'physical' body.
3,500 M/B.C.	Plants 3	God creates plants. Plants make oxygen and food for people.
4,600 M/B.C.	Earth 2	God gives people a place to live while in a physical body.
5,000 Million B.C. (Before Christ)	Sun Stars Stars	God creates the sun giving people light and warmth.

Directions

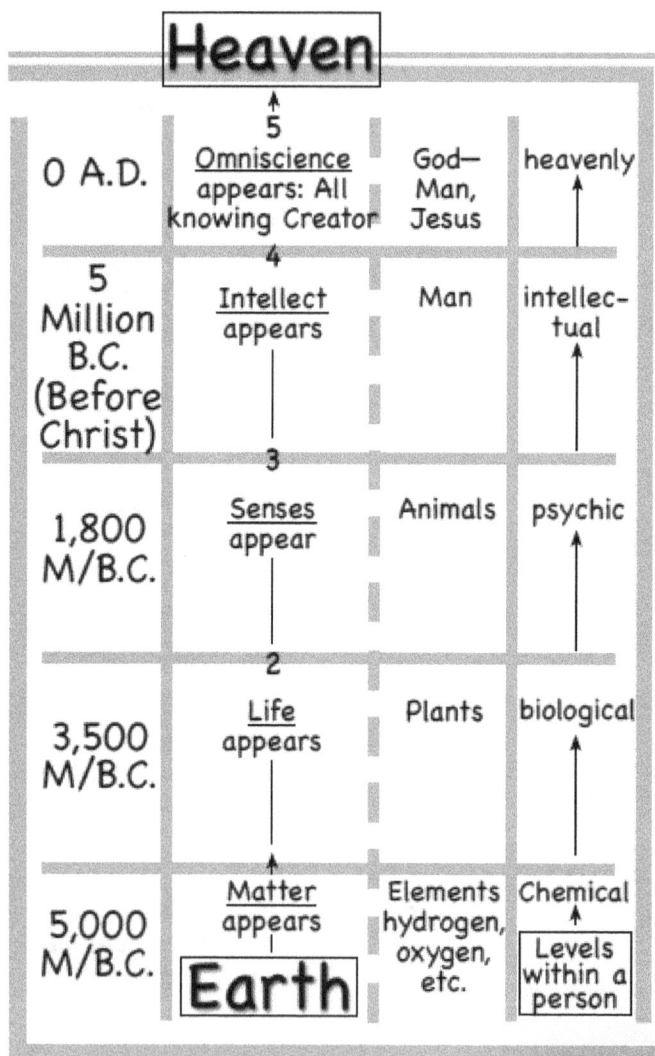

Heaven			
0 A.D.	5 <u>Omniscience</u> appears: All knowing Creator	God— Man, Jesus	heavenly
5 Million B.C. (Before Christ)	4 <u>Intellect</u> appears	Man	intellec- tual
1,800 M/B.C.	3 <u>Senses</u> appear	Animals	psychic
3,500 M/B.C.	2 <u>Life</u> appears	Plants	biological
5,000 M/B.C.	<u>Matter</u> appears **Earth**	Elements hydrogen, oxygen, etc.	Chemical Levels within a person

Plants

Plants make oxygen and food for us.

Heaven			
135 Million Before Christ	into air ↑	Flowering Trees	Seeds are inside fruit. Are carried aloft by birds and in the wind
350 M/B.C.		Evergreen Trees / dry land	Naked seeds. Grow high into the air.
400M	onto land ↑	Ferns / wet land	True <u>roots</u>, stem & leaves. Larger in size.
1,000 M		Moss	<u>No roots</u> stem or leaves. Small in size.
2,000 M	Water ↑	Green Algae X100	Plant body is <u>many</u> cells. Food reserve is starch.
3,500 M/B.C.		Blue-Green Algae X300	Plant body is <u>1-celled</u>. No reserve food.
Sea			

Leaves

Summer ➞	Fall ➞	Winter ➞	Spring
full life	fall in life	death	new life
June-Aug	Sept-Nov	Dec-Feb	Mar-May
Leaves on trees.	Leaves turn brown and fall off.	Trees are bare. Appear dead, buried in snow.	Leaves return! Tree life is re-newed.
75°F	50°F	25°F	50°F
Young person	Older person	Person at death	Risen Person

Resurrection

Grafted

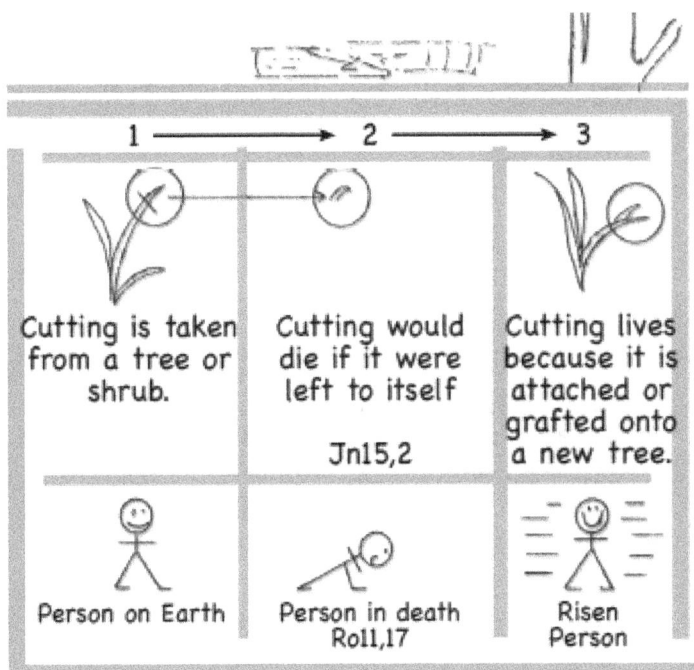

1 →	2 →	3
Cutting is taken from a tree or shrub.	Cutting would die if it were left to itself

Jn15,2 | Cutting lives because it is attached or grafted onto a new tree. |
| Person on Earth | Person in death
Ro11,17 | Risen Person |

Trees

Summer ⟶	Fall ⟶	Winter ⟶	Spring
full life	fall in life	death	new life
Seed is unplanted. Mt13,31	Seed falls upon the Earth.	Seed is 'buried', transforming.	Seed becomes a mature tree.
Young person flying high	Older person down to Earth	Person at death, buried in the Earth	Risen Person, with a new body

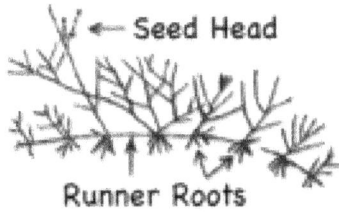

Grass

Seed Head

Runner Roots

Summer →	Fall →	Winter →	Spring
full life	fall in life	death	new life
June-Aug	Sept-Nov	Dec-Feb	Mar-May
75°F	50°F	25°F	50°F
Grass grows tall. Mt14,19	Grass turns brown. Stalks above ground die.	Grass is buried under snow. Underground roots remain alive.	Grass returns. Shows renewed life.
Young person	Older person	Person at death	Risen Person

Resurrection

Seeds

Summer →	Fall →	Winter →	Spring
Apple blossoms	Male cell 'falls' onto stigma and grows 'downward' to fertilize female egg cell	New apple seeds are protected in "fruit' (casket) during winter	Apple seed grows 'up' into a 'new apple tree'
flower pedals	stigma → egg / fallen flower pedals	seed / fruit is 'casket' for apple seeds	Resurrection
Young person	Older person	Casket / Person at death	Risen Person

Seeds & People

A corn plant rises up to become
only a corn plant. A rose seed rises up
to become only a rose plant. In the same way,
a person rises up to become only himself
or herself. We do not rise up to become
someone else or something else.

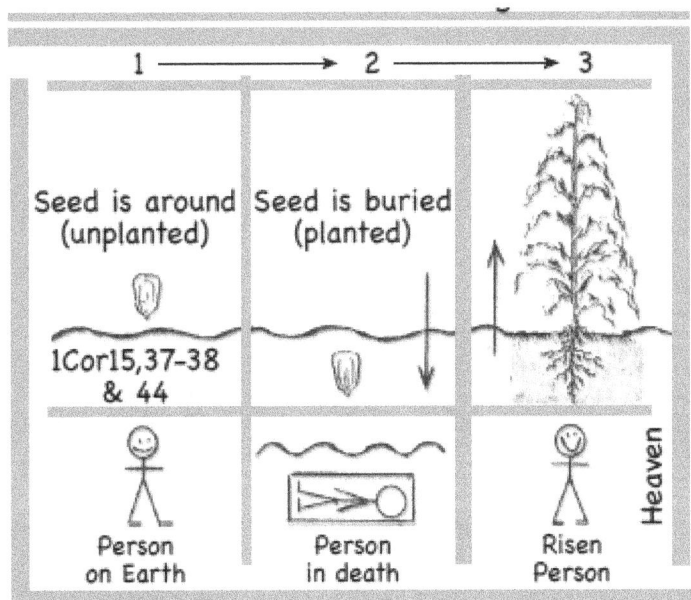

1 ——————→	2 ——————→	3
Seed is around (unplanted)	Seed is buried (planted)	
1Cor15,37-38 & 44		
Person on Earth	Person in death	Risen Person

Heaven

Rising Bread

Yeast is a living thing.
It feeds on sugar in wheat flour.
While growing, yeast gives off carbon dioxide gas (CO_2).
This gas makes bread dough rise.
After the dough has risen it is baked.

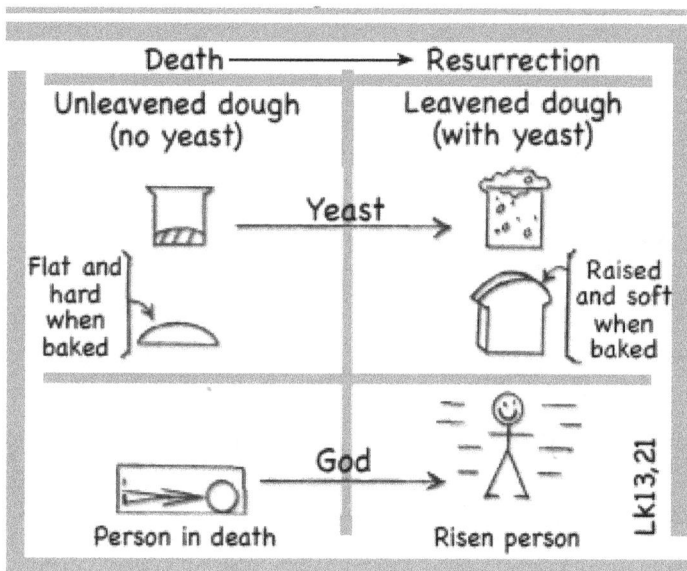

Death ——————→ Resurrection

Unleavened dough (no yeast)	Leavened dough (with yeast)

Yeast

Flat and hard when baked

Raised and soft when baked

God

Person in death   Risen person

Lk13, 21

Chapter 3

Signs of Risen Life
in Animals

God's Care

Future	**Heaven**	God creates heaven as a home for his 'spirit-ed' people.
5 M/B.C.	People 5	God creates a 'spirit' for people. People are spirits within a body.
1,800 M/B.C.	Animals 4	God creates animals. Through the history of animals people receive a 'physical' body.
3,500 M/B.C.	Plants 3	God creates plants. Plants make oxygen and food for people.
4,600 M/B.C.	Earth 2	God gives people a place to live while in a physical body.
5,000 Million B.C. (Before Christ)	Sun Stars Stars	God creates the sun giving people light and warmth.

Directions

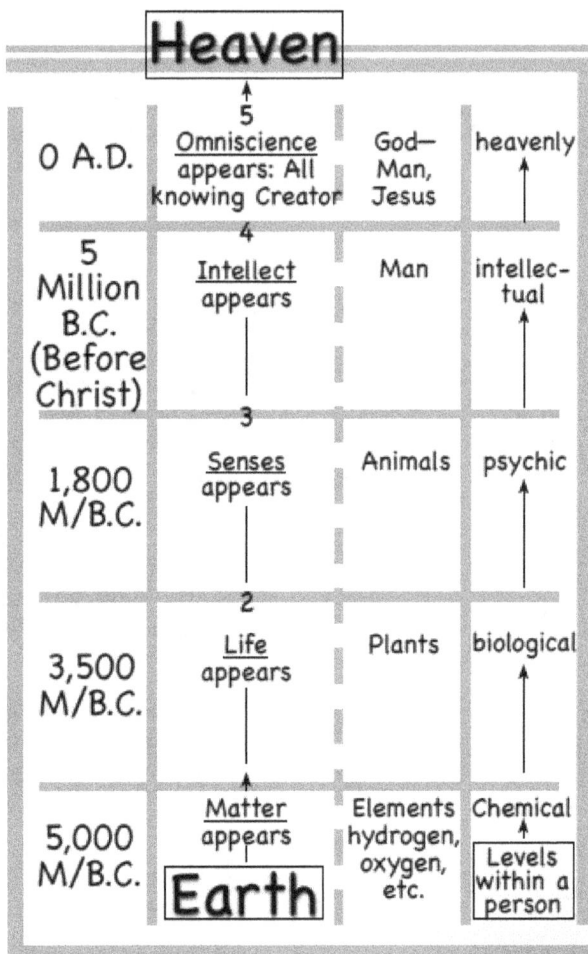

	Heaven		
0 A.D.	5 <u>Omniscience</u> appears: All knowing Creator	God— Man, Jesus	heavenly
5 Million B.C. (Before Christ)	4 <u>Intellect</u> appears	Man	intellec- tual
1,800 M/B.C.	3 <u>Senses</u> appears	Animals	psychic
3,500 M/B.C.	2 <u>Life</u> appears	Plants	biological
5,000 M/B.C.	<u>Matter</u> appears **Earth**	Elements hydrogen, oxygen, etc.	Chemical Levels within a person

Creatures

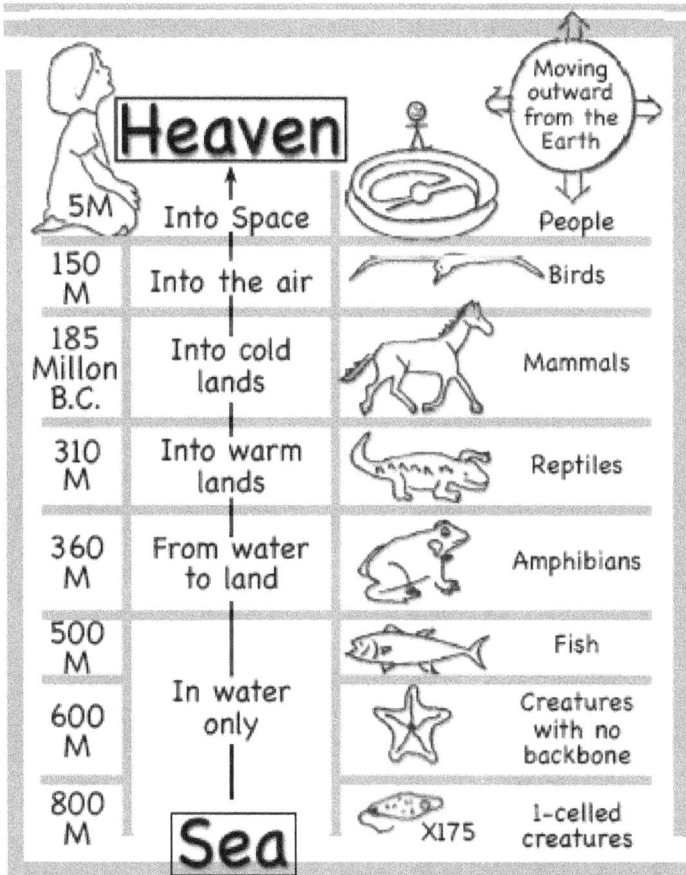

5M	Into Space	People
150 M	Into the air	Birds
185 Millon B.C.	Into cold lands	Mammals
310 M	Into warm lands	Reptiles
360 M	From water to land	Amphibians
500 M	In water only	Fish
600 M		Creatures with no backbone
800 M		1-celled creatures

Heaven

Moving outward from the Earth

Sea

X175

Sponge

There are 10,000 different kinds of sponges
on Earth. Each has 2 bodies.

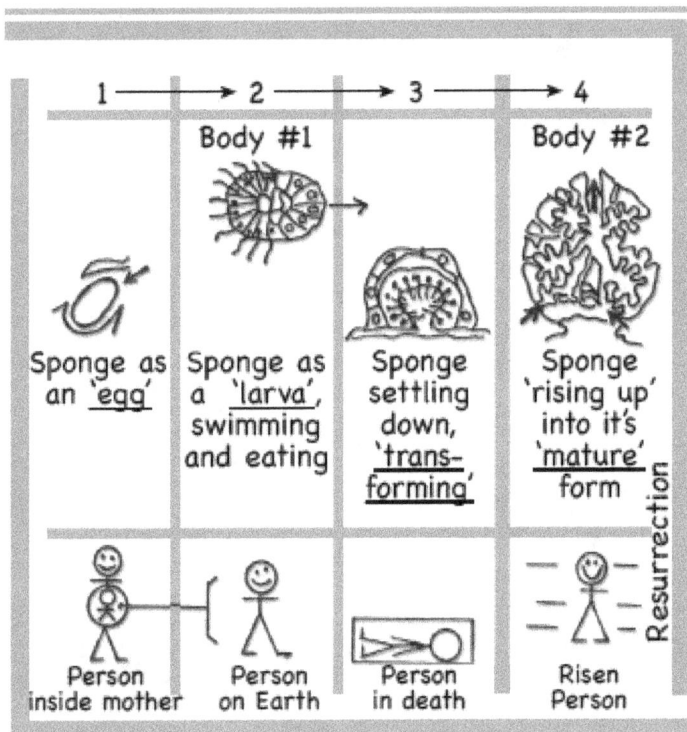

1 →	2 →	3 →	4
	Body #1		Body #2
Sponge as an 'egg'	Sponge as a 'larva', swimming and eating	Sponge settling down, 'trans-forming'	Sponge 'rising up' into it's 'mature' form
Person inside mother	Person on Earth	Person in death	Risen Person

Resurrection

Jellyfish

There are 200 different kinds of jellyfish.
Each has 2 bodies.

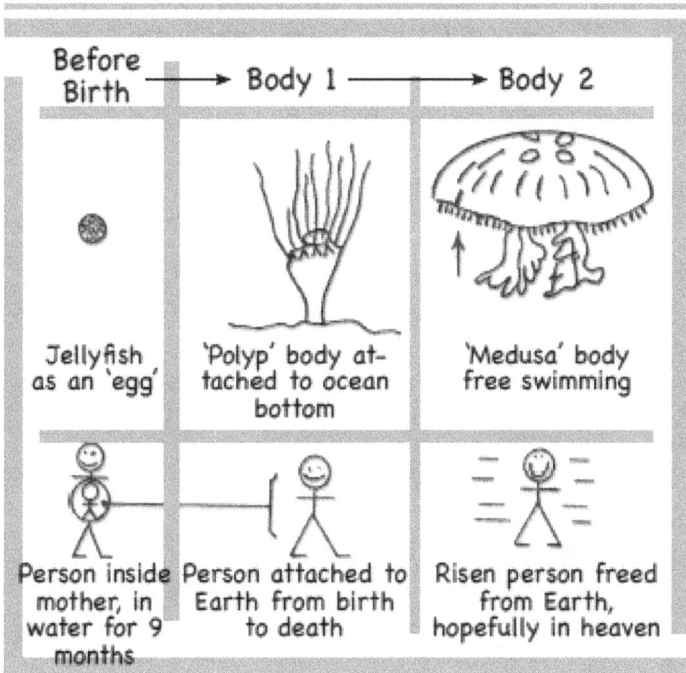

Before Birth	→ Body 1 →	Body 2
Jellyfish as an 'egg'	'Polyp' body attached to ocean bottom	'Medusa' body free swimming
Person inside mother, in water for 9 months	Person attached to Earth from birth to death	Risen person freed from Earth, hopefully in heaven

Starfish Family

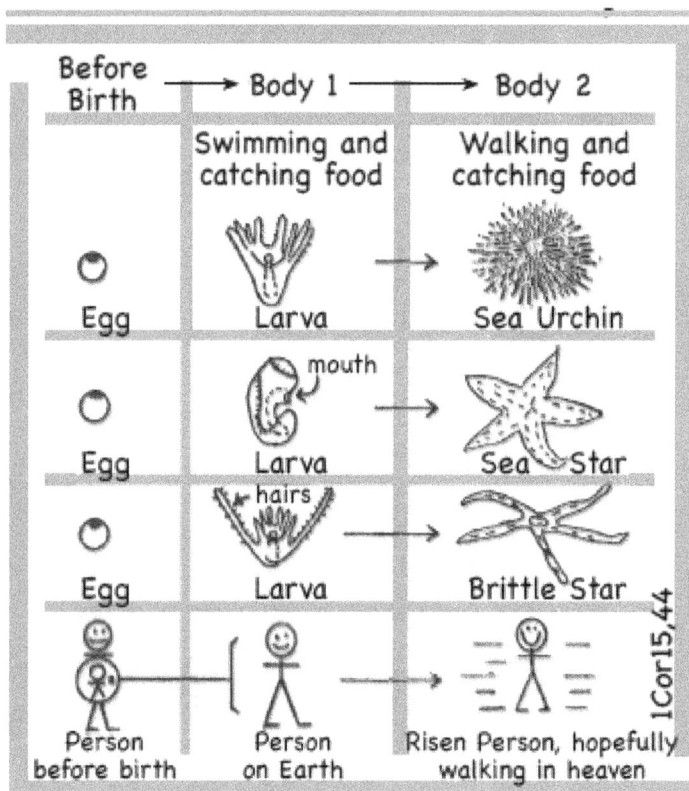

Before Birth	→ Body 1	→ Body 2
	Swimming and catching food	Walking and catching food
Egg	Larva	Sea Urchin
Egg	Larva mouth	Sea Star
Egg	Larva hairs	Brittle Star
Person before birth	Person on Earth	Risen Person, hopefully walking in heaven

1Cor15,44

Regeneration

Some creatures can re-grow
a whole new body from almost nothing.
This is a sign of our own resurrection.

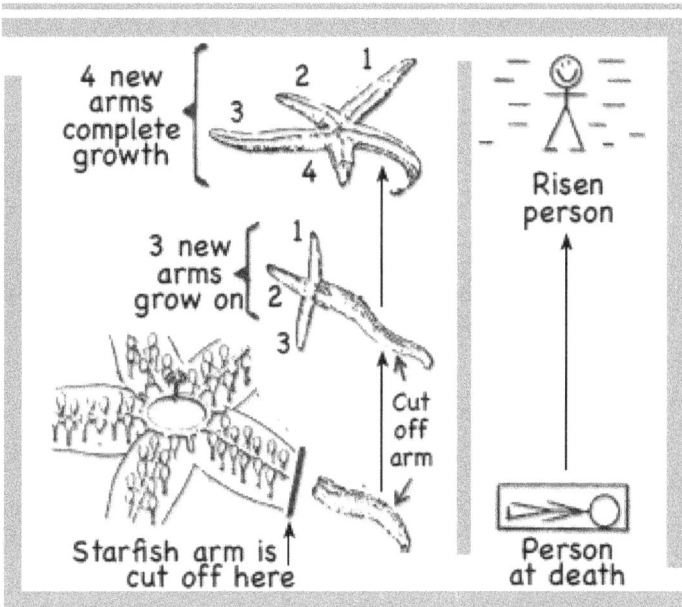

4 new
arms
complete
growth

Risen
person

3 new
arms
grow on

Cut
off
arm

Starfish arm is
cut off here

Person
at death

Bivalves

Bivalves are sea creatures having a
2-piece shell. There are 20,000 specific kinds.
They include clams, muscles, oysters and
scallops. Each has 2 bodies.

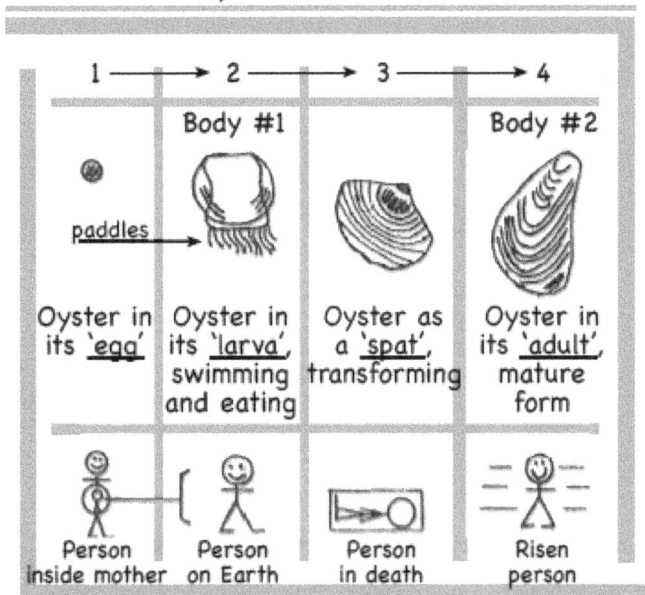

1 →	2 →	3 →	4
	Body #1		Body #2
paddles			
Oyster in its 'egg'	Oyster in its 'larva', swimming and eating	Oyster as a 'spat', transforming	Oyster in its 'adult', mature form
Person inside mother	Person on Earth	Person in death	Risen person

Butterfly

There are 100,000 different kinds of butterflies on Earth. Each butterfly has 2 bodies: a walking body and a flying body.

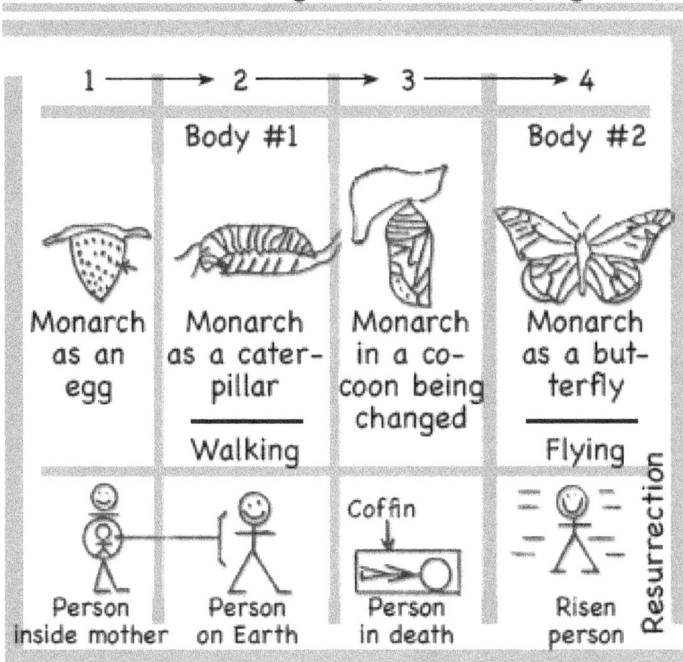

1 →	2 →	3 →	4
	Body #1		Body #2
Monarch as an egg	Monarch as a caterpillar	Monarch in a cocoon being changed	Monarch as a butterfly
	Walking		Flying
Person inside mother	Person on Earth	Coffin Person in death	Risen person

Resurrection

Honeybee

There 12,000 different kinds of
Bees on Earth. Each has 2 bodies:
a grub body and a flying body.

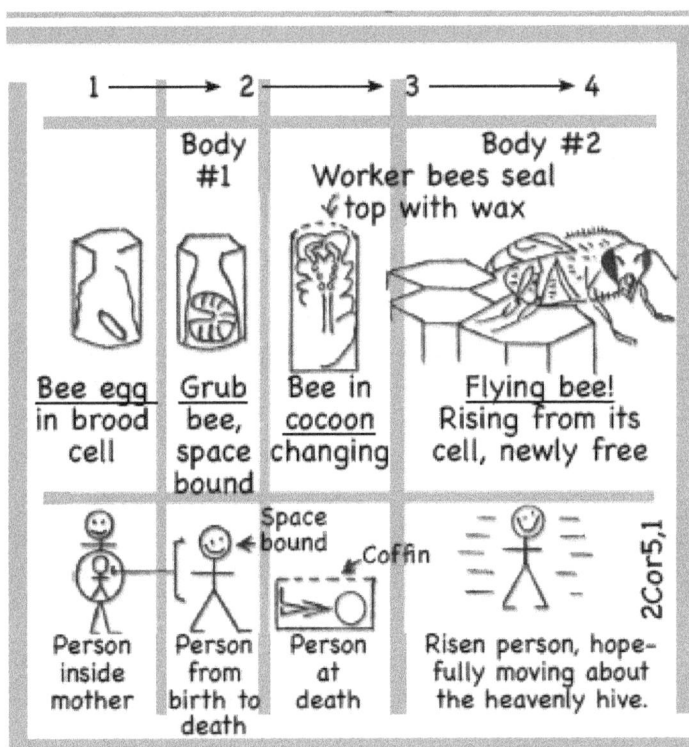

Silkworm

The silkworm spins a cocoon made of a single 1000 foot thread of silk in 3 days.

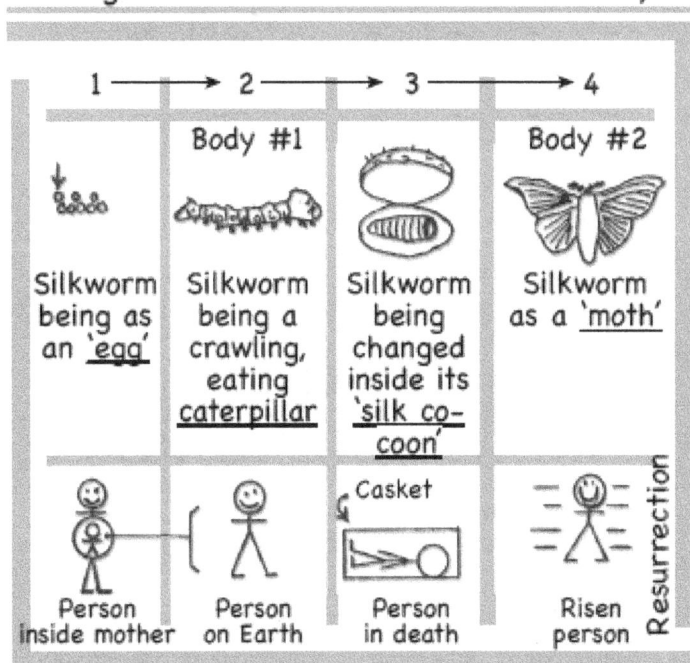

1 →	2 →	3 →	4
	Body #1		Body #2
Silkworm being as an 'egg'	Silkworm being a crawling, eating caterpillar	Silkworm being changed inside its 'silk cocoon'	Silkworm as a 'moth'
Person inside mother	Person on Earth	Casket / Person in death	Risen person

Resurrection

Ladybug

There are 300,000 kinds of beetles
on Earth. The ladybug is one kind.
All beetles have a walking body
first and then, lastly, a flying body.

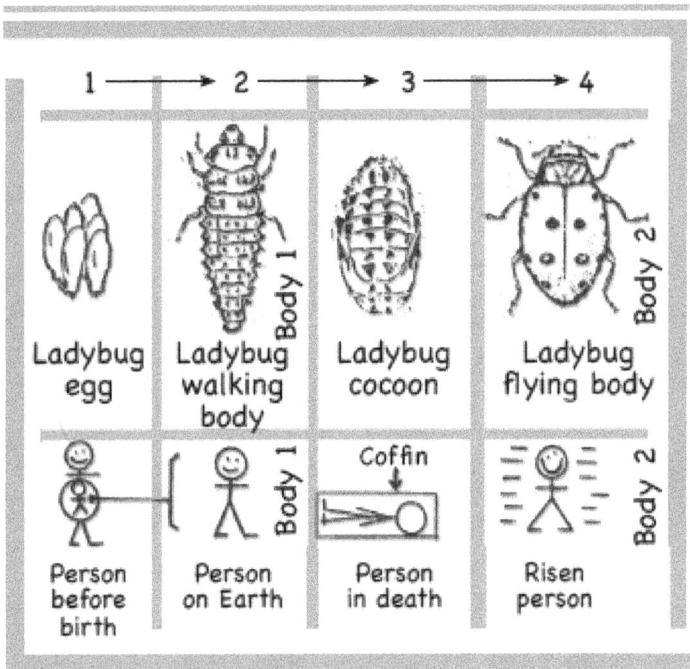

1 →	2 →	3 →	4
Ladybug egg	Ladybug walking body — Body 1	Ladybug cocoon	Ladybug flying body — Body 2
Person before birth	Person on Earth — Body 1	Coffin / Person in death	Risen person — Body 2

Fly

There are 100,000 different kinds
of flies on Earth. Each has 2 bodies:
a "maggot" body and a 'fly'ing body.

1 →	2 →	3 →	4
	Body #1		**Body #2**
Fly as an 'egg'	Fly as a crawling, eating 'maggot'	Fly in a 'cocoon case' being changed	Fly as a 'fly'
Person inside mother	Person on Earth	Casket Person in death	Risen person

Job 25,6

Resurrection

Hibernating
Earthworm

Summer →	Fall →	→ Winter →	Spring
June-Aug	Sept-Nov	Dec-Feb	Mar-May
75°F		25°F snow	
Worm is active on top of soil in grass	Worm is active within soil, near top	Worm sleeps 3 feet under ground	Worm is back in top soil, or in grass
Young person, on top of things	Older person, slowly going under	Person at death	Risen person back on top

Resurrection

Earthworm

The earthworm can regrow it's body parts. This power comes from God. At our death God will use this power on us.

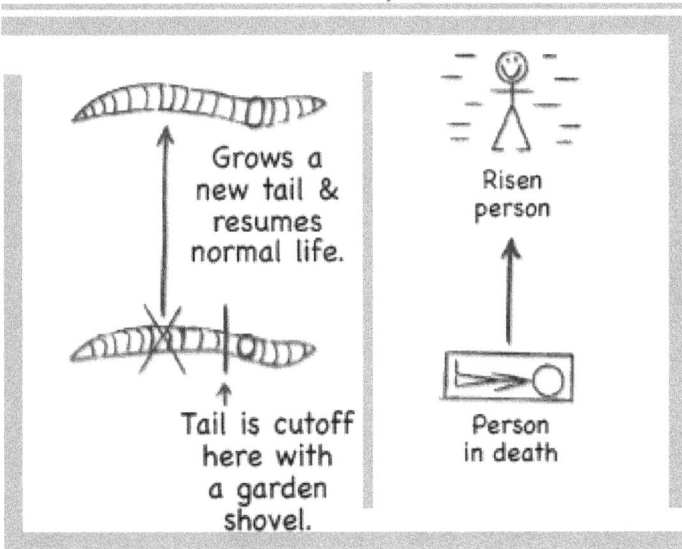

Grows a new tail & resumes normal life.

Tail is cutoff here with a garden shovel.

Risen person

Person in death

Bullhead Catfish

Summer →	Fall →	Winter →	Spring
Swims around in full strength	Is less active as water temperature drops	Sleeps in mud at the lake bottom	Rises up to new life
Young person	Older person	Person at death	Risen person

Lake bottom

Resurrection

Frog

There are 800 different kinds of frogs
on Earth. Every frog has 2 bodies.

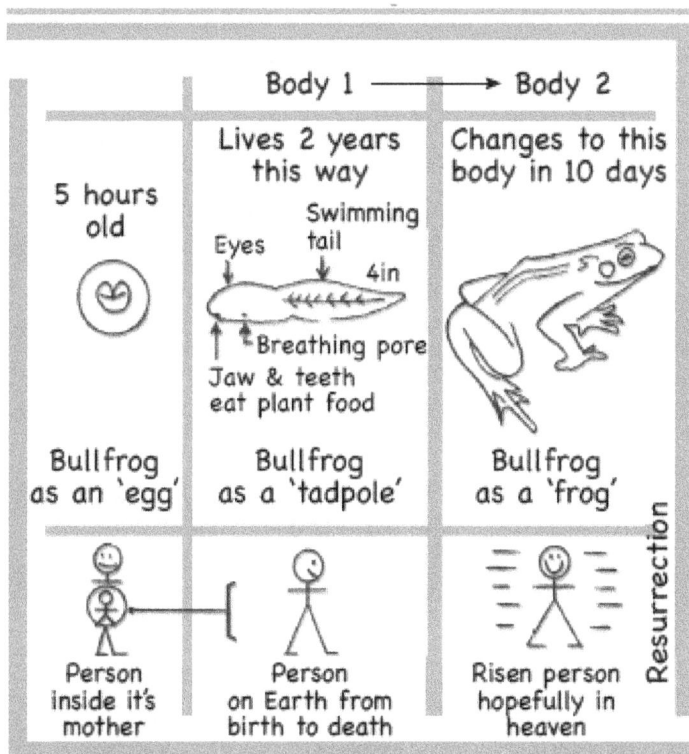

	Body 1 ⟶ Body 2	
5 hours old ⊛	**Lives 2 years this way** Swimming tail / Eyes / 4in / Breathing pore / Jaw & teeth eat plant food	**Changes to this body in 10 days**
Bullfrog as an 'egg'	Bullfrog as a 'tadpole'	Bullfrog as a 'frog'
Person inside it's mother	Person on Earth from birth to death	Risen person hopefully in heaven / Resurrection

Wood Frog

Freezes in winter. Its heart and breathing stop for months.

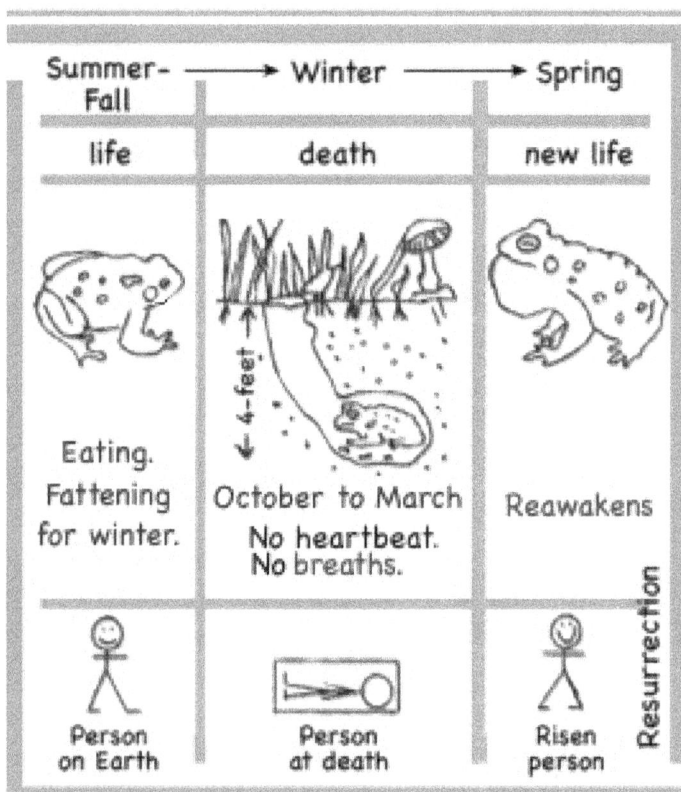

Summer-Fall ⟶	Winter ⟶	Spring
life	death	new life
Eating. Fattening for winter.	October to March No heartbeat. No breaths.	Reawakens
Person on Earth	Person at death	Risen person

Resurrection

Turtles

Summer →	Fall →	Winter →	Spring
full life	fall in life	death	new life
Eating! Putting on weight to hibernate	Slowing down as the temperature 'falls'	Sleeping near death. Heart- beat:5/min Breaths: 1/min	Rising as the ground warms
75°F	50°F	25°F	50°F
June-Aug	Sept-Nov	Dec-Feb	March-May
Young person	Older person	Person at death	Risen person

Birds

Death →	Resurrection →	Heaven
Incubating egg	Hatching egg	Bird in flight
Person at death in a tomb. Space and time bound.	Person at the time of resurrection.	Risen person hopefully in heaven.

Mt 23,37 · Lk12,16 · Proverbs23,5

Bird Migration

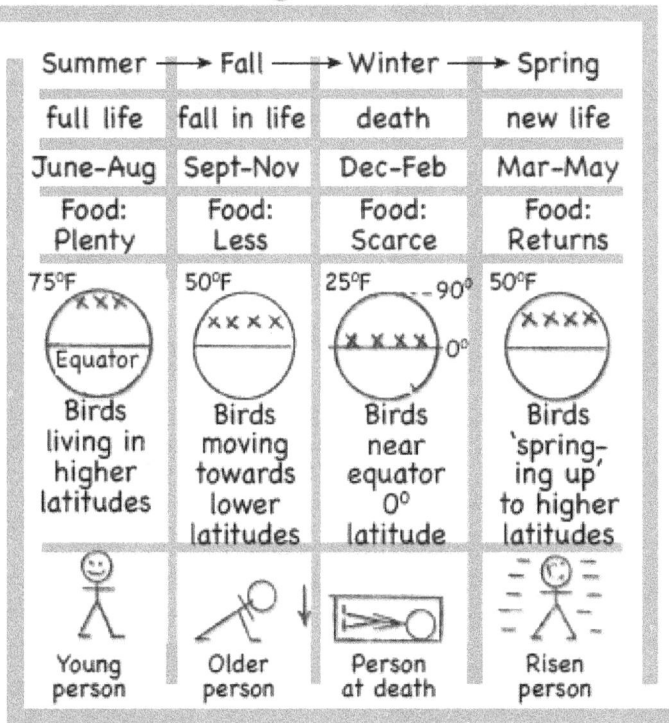

Summer →	Fall →	Winter →	Spring
full life	fall in life	death	new life
June-Aug	Sept-Nov	Dec-Feb	Mar-May
Food: Plenty	Food: Less	Food: Scarce	Food: Returns
75°F	50°F	25°F	50°F
Birds living in higher latitudes	Birds moving towards lower latitudes	Birds near equator 0° latitude	Birds 'springing up' to higher latitudes
Young person	Older person	Person at death	Risen person

Chipmunk

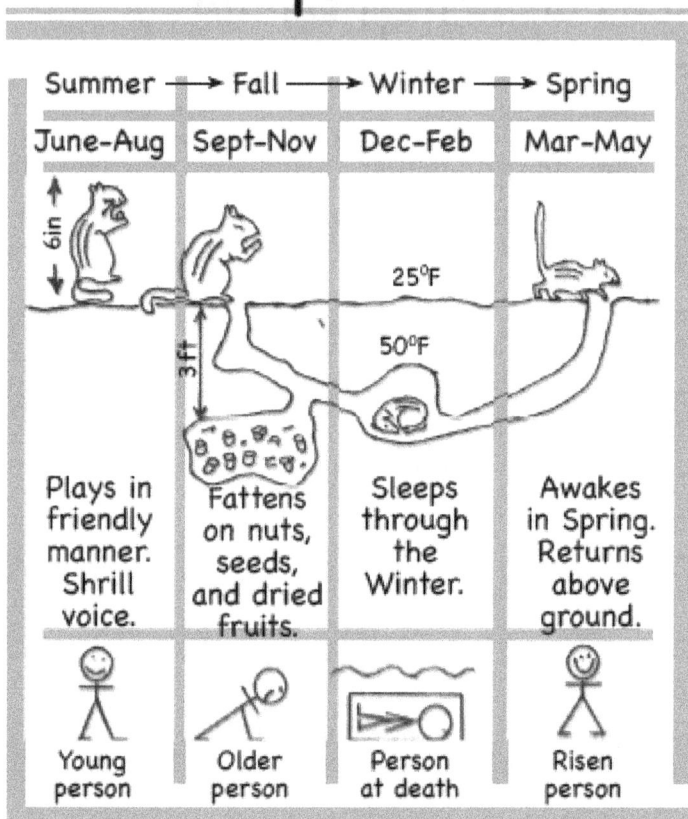

Summer →	Fall →	Winter →	Spring
June-Aug	Sept-Nov	Dec-Feb	Mar-May
Plays in friendly manner. Shrill voice.	Fattens on nuts, seeds, and dried fruits.	Sleeps through the Winter.	Awakes in Spring. Returns above ground.
Young person	Older person	Person at death	Risen person

Black Bear

Summer →	Fall ──→	Winter ──→	Spring
full life	fall in life	death	new life
June-Aug	Sept-Nov	Dec-Feb	Mar-May
Eating nuts and insects	Fattening for Winter	Sleeping from December to March	Waking 'up'
Young person	Older person	Person at death	Risen person

Resurrection

Polar Bear

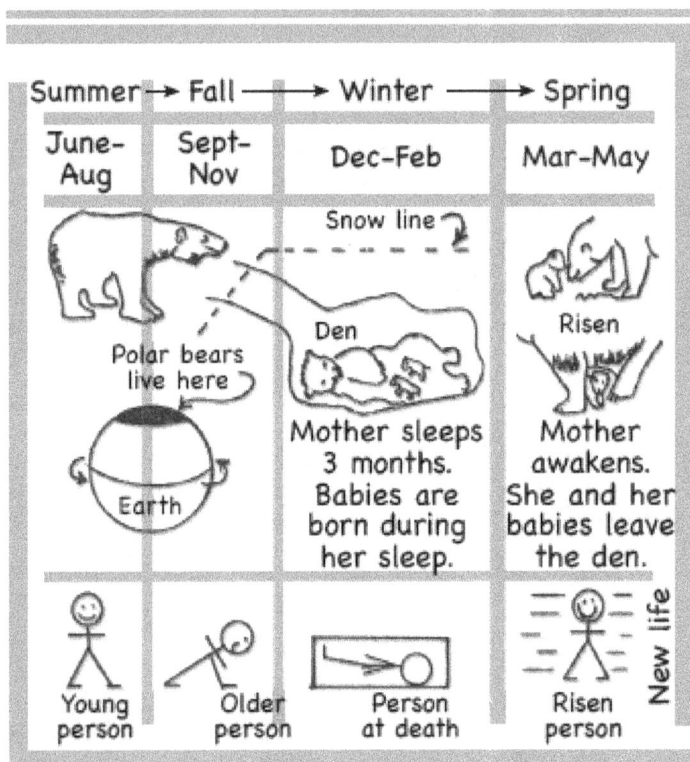

Chapter 4

Signs of Risen Life in the Human Body

God's Care

Future	**Heaven**	God creates heaven as a home for his 'spirit-ed' people.
5 M/B.C.	People 5	God creates a 'spirit' for people. People are spirits within a body.
1,800 M/B.C.	Animals 4	God creates animals. Through the history of animals people receive a 'physical' body.
3,500 M/B.C.	Plants 3	God creates plants. Plants make oxygen and food for people.
4,600 M/B.C.	Earth 2	God gives people a place to live while in a physical body.
5,000 Million B.C. (Before Christ)	Sun Stars Stars	God creates the sun giving people light and warmth.

Directions

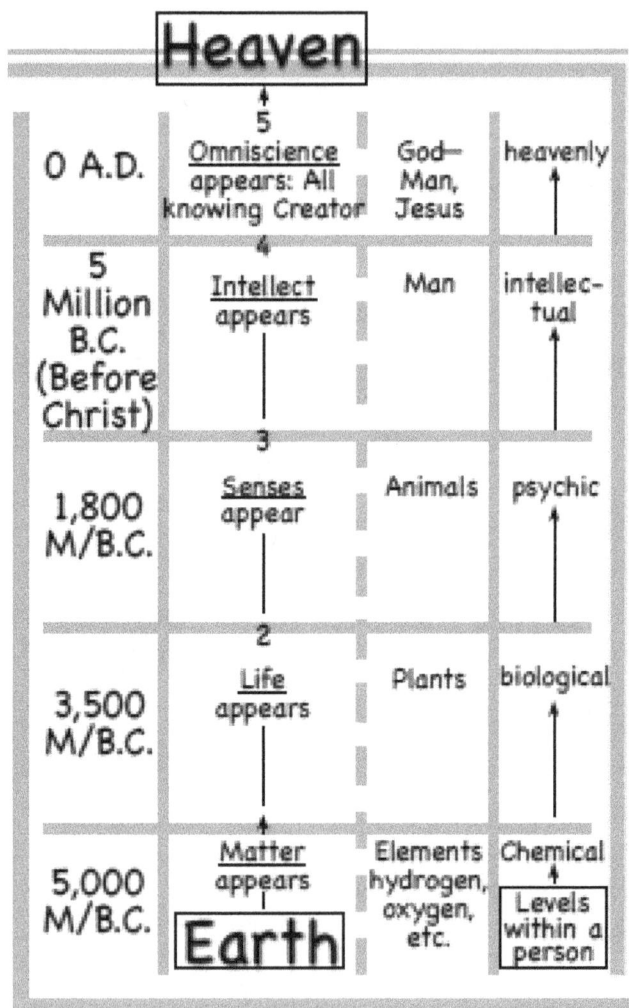

	Heaven		
	↑ 5		
0 A.D.	<u>Omniscience</u> appears: All knowing Creator	God— Man, Jesus	heavenly ↑
5 Million B.C. (Before Christ)	4 <u>Intellect</u> appears	Man	intellec- tual ↑
1,800 M/B.C.	3 <u>Senses</u> appear	Animals	psychic ↑
3,500 M/B.C.	2 <u>Life</u> appears	Plants	biological ↑
5,000 M/B.C.	1 <u>Matter</u> appears **Earth**	Elements hydrogen, oxygen, etc.	Chemical ↑ Levels within a person

Birth

Skin

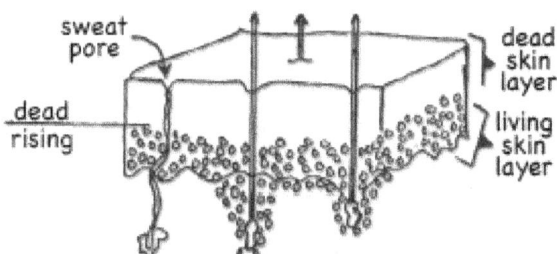

sweat pore

dead rising

dead skin layer

living skin layer

Skin has an outer and an inner side.
The outer side that people see is dead.
The inner side is living and growing. The
outer side is a sign of our coming death.
The inner living side is a sign of our
coming resurrection.
Like our death the outer side
can be seen. Like our resurrection the inner
living side is not seen. We carry these signs of
death and resurrection with us at all times.

Hair

Every hair we see on a person's body is dead. Only the root of each hair is alive. The root hair is 'buried' in the scalp underneath the outer layer of skin. The dead hair that we see on the outside is a sign of our coming death. The living hair root underneath that we cannot see is a sign of our coming resurrection.

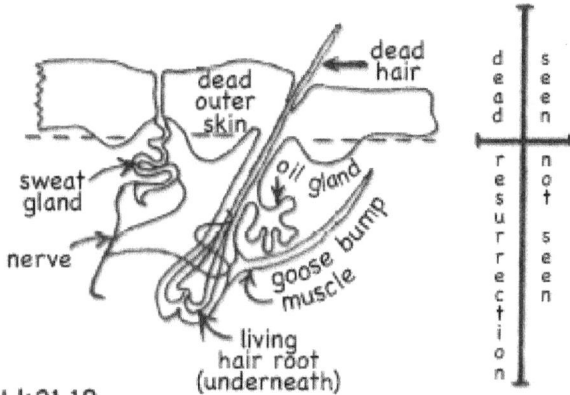

dead hair

dead outer skin

sweat gland

oil gland

nerve

goose bump muscle

living hair root (underneath)

Lk 21,18

dead / seen

resurrection / not seen

Hunger

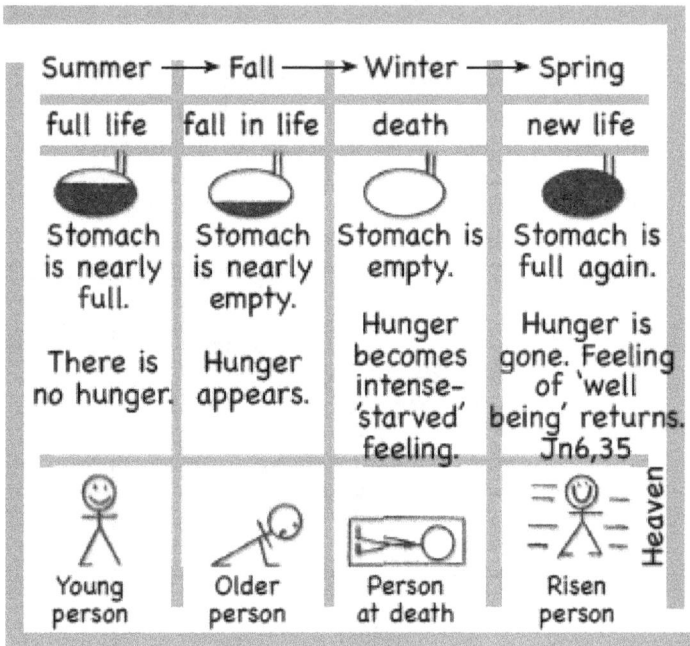

Summer →	Fall →	Winter →	Spring
full life	fall in life	death	new life
Stomach is nearly full.

There is no hunger. | Stomach is nearly empty.

Hunger appears. | Stomach is empty.

Hunger becomes intense- 'starved' feeling. | Stomach is full again.

Hunger is gone. Feeling of 'well being' returns. Jn6,35 |
| Young person | Older person | Person at death | Risen person |

Sight

Not to see is a sign of death.
Restored sight is a sign of resurrection.

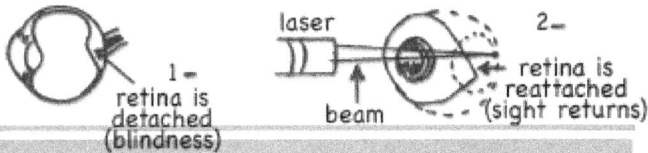

Summer →	Fall →	Winter →	Spring
Clear eyes	Clouding eyes due to cataracts	Blindness. Not able to see Mt11,5	Cataracts removed. Eyes are clear again!
Young person	Older person	Person at death	Risen person

Mk10,49

Inner Growth

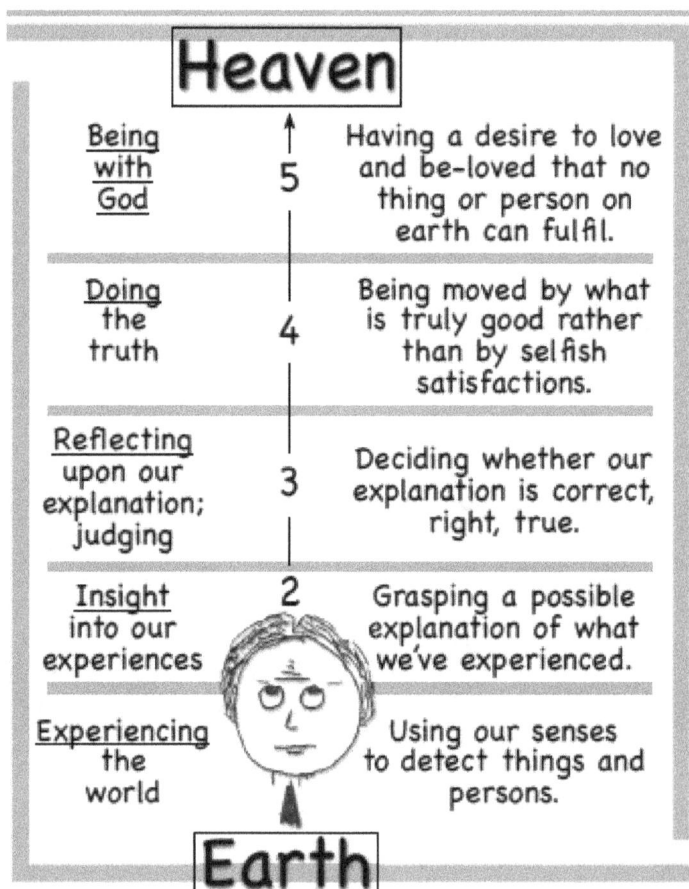

Heaven		
<u>Being with God</u>	↑ 5	Having a desire to love and be-loved that no thing or person on earth can fulfil.
<u>Doing the</u> truth	4	Being moved by what is truly good rather than by selfish satisfactions.
<u>Reflecting</u> upon our explanation; judging	3	Deciding whether our explanation is correct, right, true.
<u>Insight</u> into our experiences	2	Grasping a possible explanation of what we've experienced.
<u>Experiencing</u> the world		Using our senses to detect things and persons.
Earth		

Insight

An insight is like resurrection.
It raises one up to new life,
to new understanding.

Summer →	Fall →	Winter →	Spring
All is well. No problems.	Problem arises. Feeling puzzled.	Solution does not seem possible.	Insight occurs. Problem is solved.
Young person	Older person	Person at death	Risen person

Hypnosis

One of the signs of death is forgetting.
One of the signs of resurrection is
remembering. Hypnosis raises lost
memories back to life.

Summer →	Fall →	Winter →	Spring
Present memory	Partial memory	No memory	Full memory
A person sees or hears something	Short time later only half is remembered	Long time later nothing is remembered	Hypnosis is used to restore a person's memory
Young person	Older person	Person at death	Risen person

LK12,2

Raising People

People live longer and longer lives on earth. This is a sign of resurrection because in heaven we will live forever.

Lk 7,14 & Rev 21,4

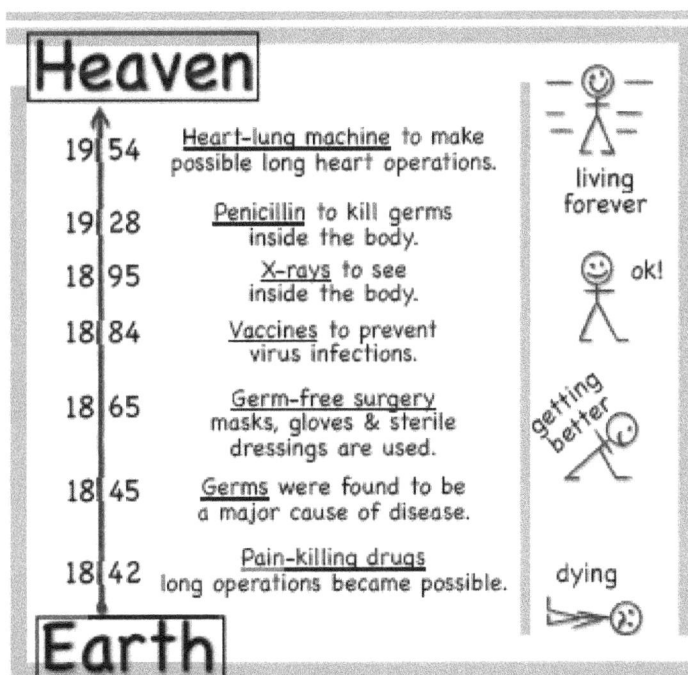

Pain

Doctors use anesthesia to stop pain. This is a sign of resurrection because in heaven there will be no more pain. Rev21,4

Summer →	Fall →	Winter →	Spring
full life	fall in life	death	new life
Feel little or no pain.	Feel some pain.	Feel great pain (from an accident, disease, or surgery).	Pain is ended by using pain-killing drugs.
Young person	Older person	Person at death	Risen person

New Body

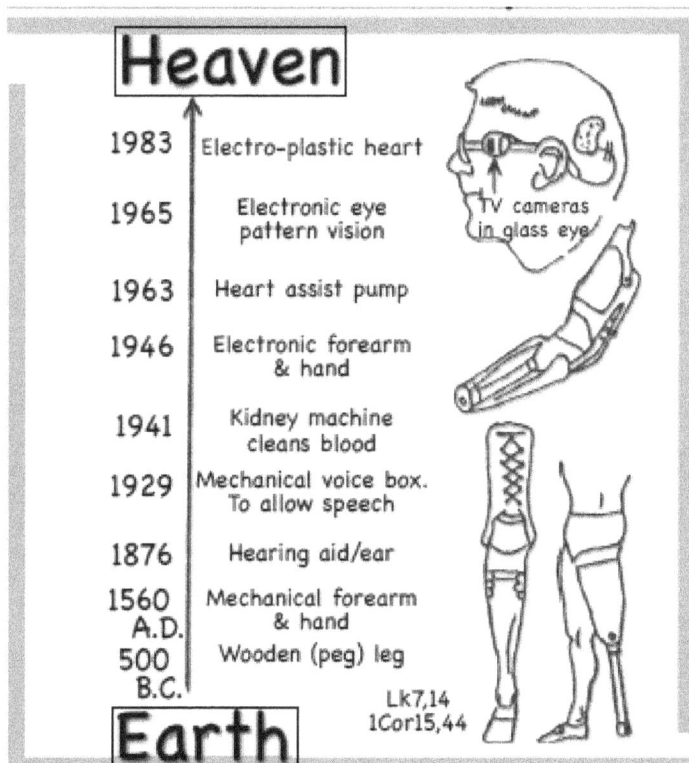

Heaven

Year	Description
1983	Electro-plastic heart
1965	Electronic eye pattern vision
1963	Heart assist pump
1946	Electronic forearm & hand
1941	Kidney machine cleans blood
1929	Mechanical voice box. To allow speech
1876	Hearing aid/ear
1560 A.D.	Mechanical forearm & hand
500 B.C.	Wooden (peg) leg

TV cameras in glass eye

Lk7,14
1Cor15,44

Earth

Extending Body

Machines extend a person's body. They make it
more powerful and less bound in space.
They are signs of our new body
—hopefully in heaven.

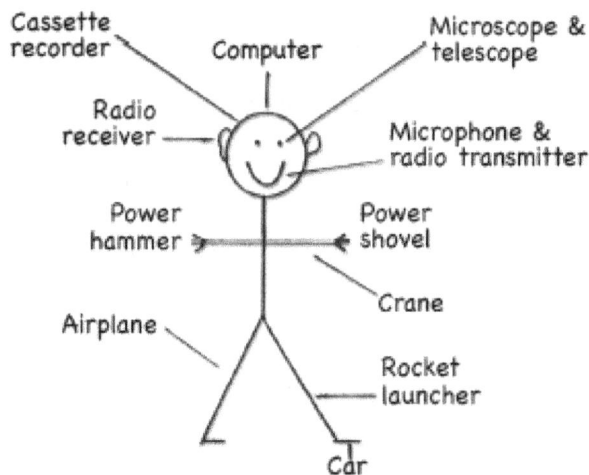

Cassette
recorder

Computer

Microscope &
telescope

Radio
receiver

Microphone &
radio transmitter

Power
hammer

Power
shovel

Crane

Airplane

Rocket
launcher

Car

Chapter 5

Signs of Risen Life in Human Behavior

Needs

Heaven		
5. Need to complete self	Developing one's talents to their fullest.	Will vary from person to person and be completed in heaven.
4. Esteem needs	Achievement and prestige.	As love needs are met, the need for success and prestige emerge.
3. Love needs	Affection and belonging	Friends, marriage, children, and group sharing.
2. Safety needs	Protection, order, and stability.	A safe, familiar community. Order and stability at home.
1. Body needs	Food, clothing, and shelter.	A starving man (need 1) will care little about how he is seen by others (need 4).

Earth

Sleeping-Waking

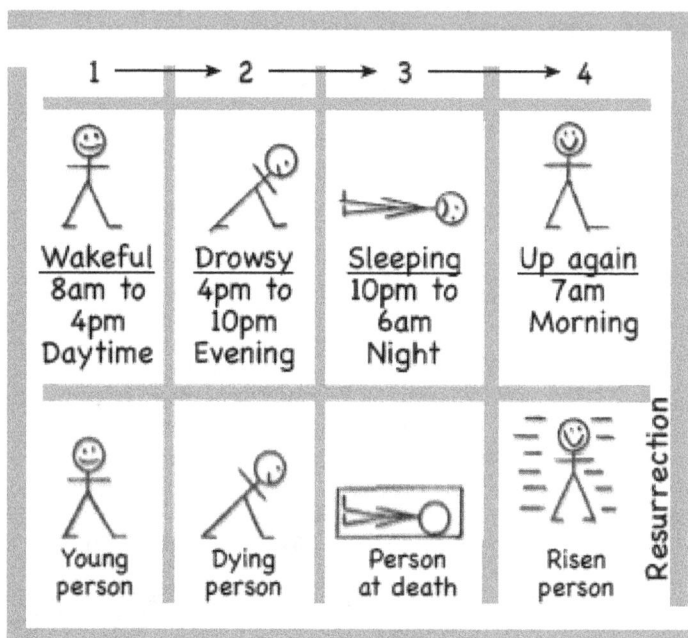

1 →	2 →	3 →	4
Wakeful 8am to 4pm Daytime	Drowsy 4pm to 10pm Evening	Sleeping 10pm to 6am Night	Up again 7am Morning
Young person	Dying person	Person at death	Risen person

Resurrection

Dreaming

We sleep.

This is a sign of our death.

At the same time, we dream. This is a sign of our resurrection. Dreaming is a special kind of life. It is a hidden life. We do not see a person's dreams.

Dreams are like a person's resurrection.

We do not usually see it either.

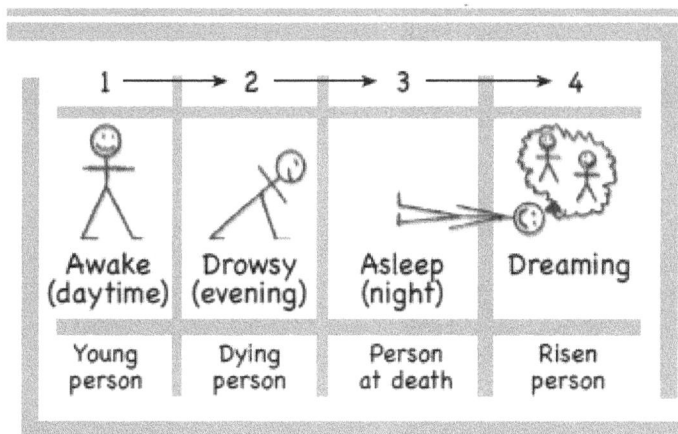

1 → 2 → 3 → 4			
Awake (daytime)	Drowsy (evening)	Asleep (night)	Dreaming
Young person	Dying person	Person at death	Risen person

Bad Dreams
in Children

Summer →	Fall →	Winter →	Spring
full life	fall in life	death	new life
Trouble-free sleep.	Bad dreams occur.	Nightmare occurs.	Mother comes. All is well.
Young person	Older person. Thoughts of death.	Person at death	Risen person. God comes. All is well.

Bad Dreams

10 P.M. →	2 A.M. →	6 A.M. →	10 A.M.
Settling down in bed.	Sleeping well 'under'.	Death-like, struggle to awake.	Mid-morning resurrection. Fully awake.
Spinning cocoon	Cocoon experience	Struggling out	Risen

Eating

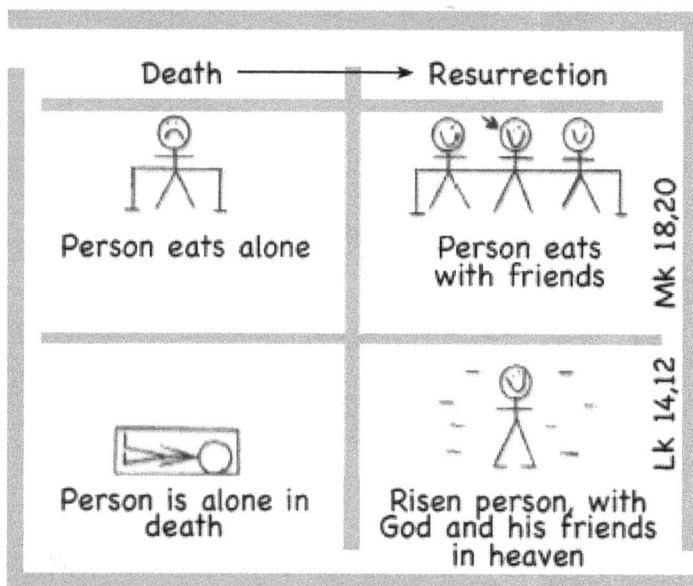

Death ——→	Resurrection	
Person eats alone	Person eats with friends	Mk 18,20
Person is alone in death	Risen person, with God and his friends in heaven	Lk 14,12

Clothes

Summer ⟶	Fall ⟶	Winter ⟶	Spring
full life	fall in life	death	new life
Our present clothes	Aging clothes	Clothes thrown out	New clothes
Young body.	Aging body.	Body dies.	New risen body.

Resurrection

Mk 5,28

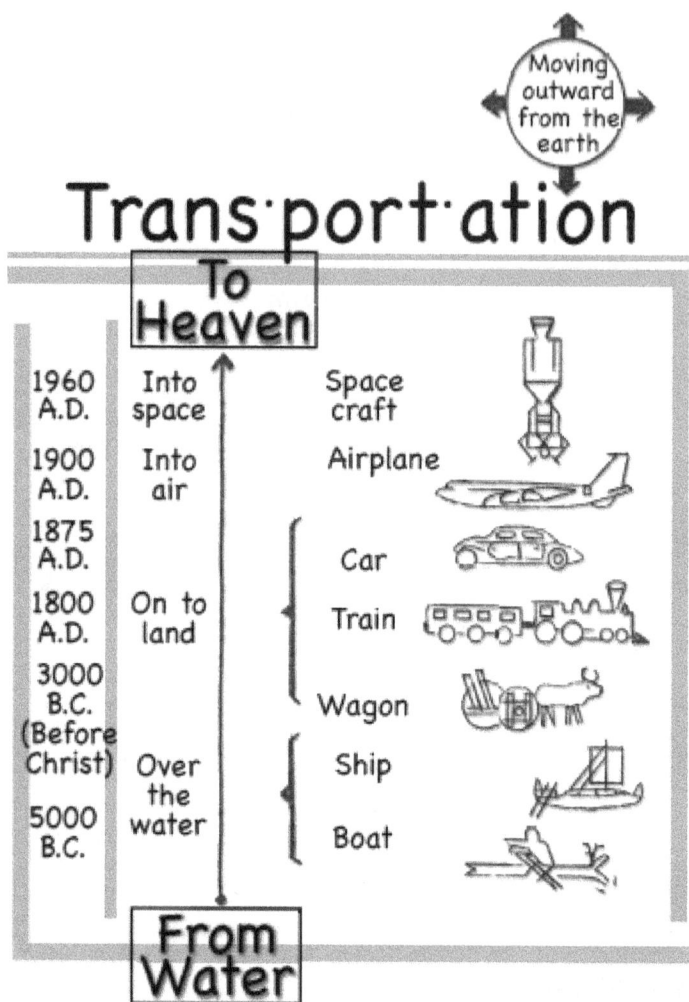

Moving outward from the earth

Trans·port·ation

To Heaven

1960 A.D.	Into space	Space craft
1900 A.D.	Into air	Airplane
1875 A.D.		Car
1800 A.D.	On to land	Train
3000 B.C. (Before Christ)		Wagon
	Over the water	Ship
5000 B.C.		Boat

From Water

Work Week

Summer →	Fall →	Winter →	Spring
full life	fall in life	death	new life
Monday and Tuesday	Wednesday and Thursday	Friday and Saturday	Sunday
Full strength, full output	Fall in strength, falling output	Tired-recoup.	Renewed strength.
Young person	Older person	Person at death	Risen person

Heaven

Practice

Practice makes a good work perfect.
Practice also increases one's freedom. Such
freedom and perfection are signs of heaven.

Summer →	Fall →	Winter →	Spring
		Psalm 150,4	
Practice makes perfect, but...	usually results in falls.	Practice may also put us "down and out" but...	eventually leads to a more perfect performance.
			Heaven
Young person	Older person	Person at death	Risen person

Habits

When a way of behaving is
repeated it becomes a habit.
There are good and bad habits.

Death		Heaven
Bad habits		**Good habits**
Mistreating parents	1.	Honoring parents
Hurting people (anger, killing)	2.	Loving people (helping them)
Dishonoring marriage (seek another mate)	3.	Honoring marriage (faithful to mate)
Steal from others	4.	Respect other's belongings
Saying untrue things (to lie)	5.	Speak truthfully
Desire what others have (envy, jealousy)	6.	Be patient with one's own life (humility)
Follow false gods (money, power, sex, Satan, wine)	7.	Follow God alone (known from the N.T., Bible)

Beauty

Heaven

beauty · great
much
more
some

Earth

Beauty is the object of our desire. We desire a beautiful body, to have beautiful food, beautiful clothing, and a beautiful house to live in. We choose to be around beautiful people, We seek beautiful things to learn, beautiful entertainment and travel. Yet our desire, here on earth, is never really satisfied. We seek more and more on earth, and when we have it, we are not really as happy as we thought we would be. This is so because we have been made by God to live in heaven. We cannot find, or create, heaven on earth.

Boredom

Life builds up when there is interest in some activity.

Summer →	Fall →	Winter →	Spring
full life	fall in life	death	new life
Things going ok.	Nothing to do.	Boredom sets in. Time stands still.	Doing something interesting. Time flies.
Young person	Older person	Person at death	Risen person

Books

A good book moves the reader
out of himself into a new world.
This is a sign of risen life.

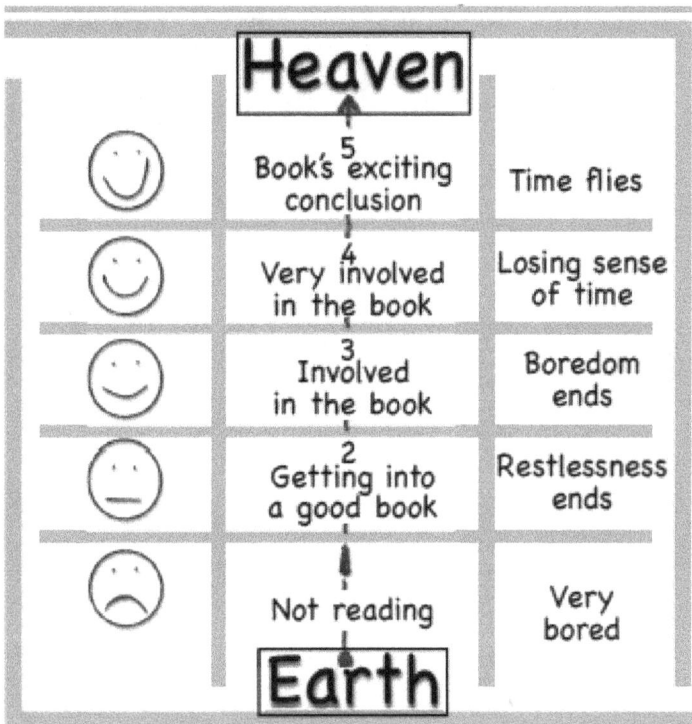

☺	Book's exciting conclusion (5)	Time flies
☺	Very involved in the book (4)	Losing sense of time
☺	Involved in the book (3)	Boredom ends
😐	Getting into a good book (2)	Restlessness ends
☹	Not reading	Very bored

Heaven ↑ ... **Earth**

Stories

Summer →	Fall →	Winter →	Spring
full life	fall in life	death	new life
Scene 1 A good person appears. There is no trouble.	Scene 2 A bad person appears. Trouble arises.	Scene 3 The bad person almost succeeds. Trouble builds to near death.	Scene 4 The good person triumphs! The bad person is overcome. Trouble ends.
Young person	Older person	Person at death	Risen person Heaven

Fall/Get Up

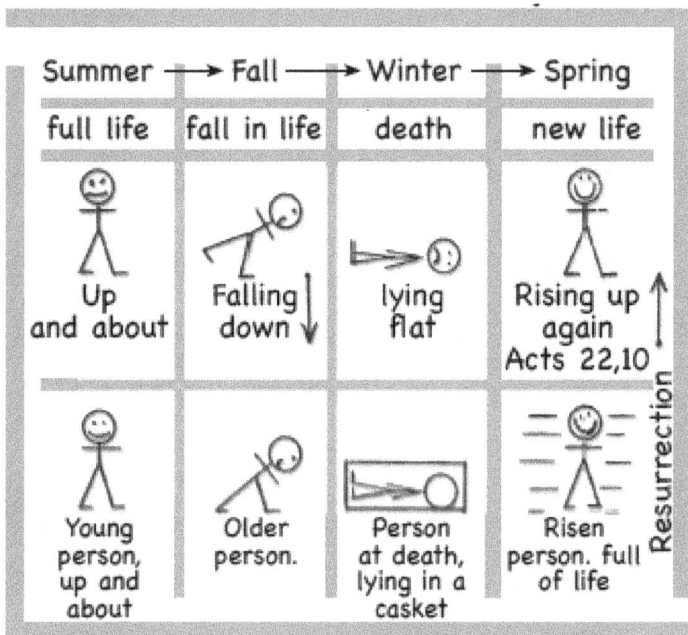

Summer →	Fall →	Winter →	Spring
full life	fall in life	death	new life
Up and about	Falling down	lying flat	Rising up again Acts 22,10
Young person, up and about	Older person.	Person at death, lying in a casket	Risen person. full of life

Sickness & Recovery

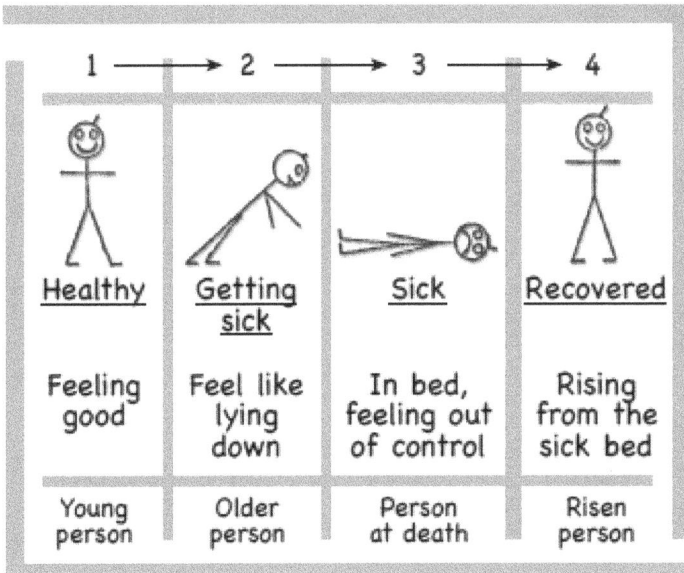

1 →	2 →	3 →	4
Healthy	Getting sick	Sick	Recovered
Feeling good	Feel like lying down	In bed, feeling out of control	Rising from the sick bed
Young person	Older person	Person at death	Risen person

Chapter 6

Signs of Risen Life
in Human Love

Relation·ships

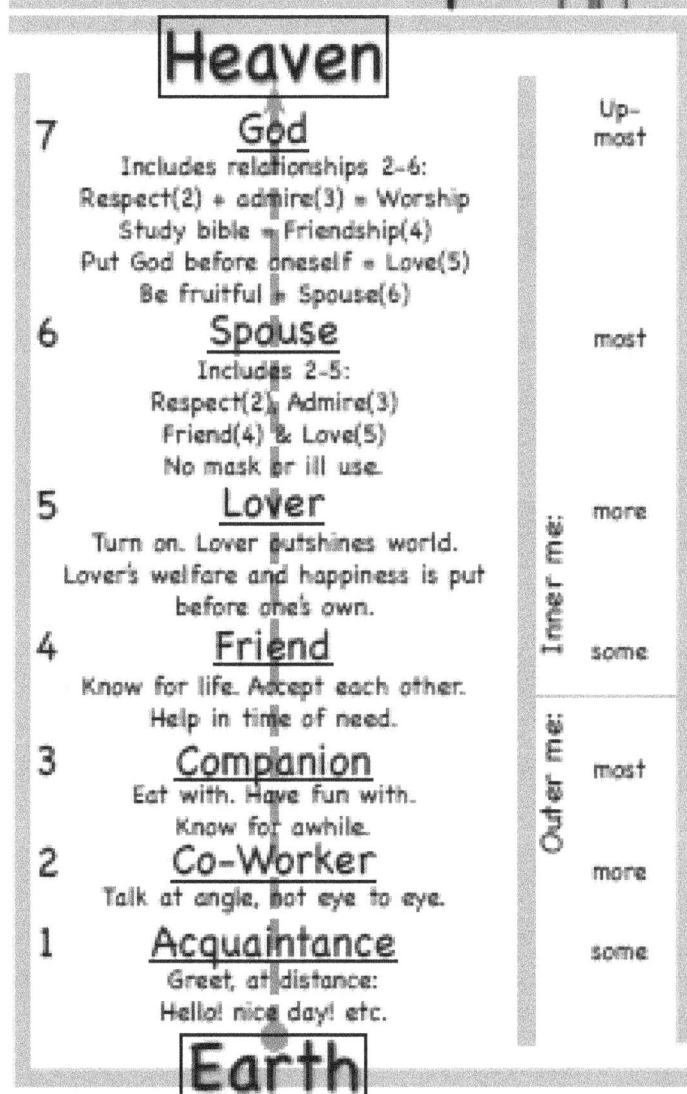

Heaven

7 God
Includes relationships 2-6:
Respect(2) + admire(3) = Worship
Study bible = Friendship(4)
Put God before oneself = Love(5)
Be fruitful = Spouse(6)

Up-most

6 Spouse
Includes 2-5:
Respect(2), Admire(3)
Friend(4) & Love(5)
No mask or ill use.

most

5 Lover
Turn on. Lover outshines world.
Lover's welfare and happiness is put
before one's own.

more

4 Friend
Know for life. Accept each other.
Help in time of need.

some

3 Companion
Eat with. Have fun with.
Know for awhile.

most

2 Co-Worker
Talk at angle, not eye to eye.

more

1 Acquaintance
Greet, at distance:
Hello! nice day! etc.

some

Inner me:

Outer me:

Earth

Love

Time stands still as in heaven.

Summer →	Fall-ing in love →	→ Winter →	Spring-ing up
age 10	age 15	age 20	age 25
The child's love comes mainly from within the family	Outside love is tried but does not take. (dating)	As the person leaves home out-side love is needed more than ever.	True out-side love appears! One's life is doubled-raised up. (marriage)
		'Lone'liness is felt (alone-ness)	
Young person	Older person	Person in death	Risen person

Dating

Loving

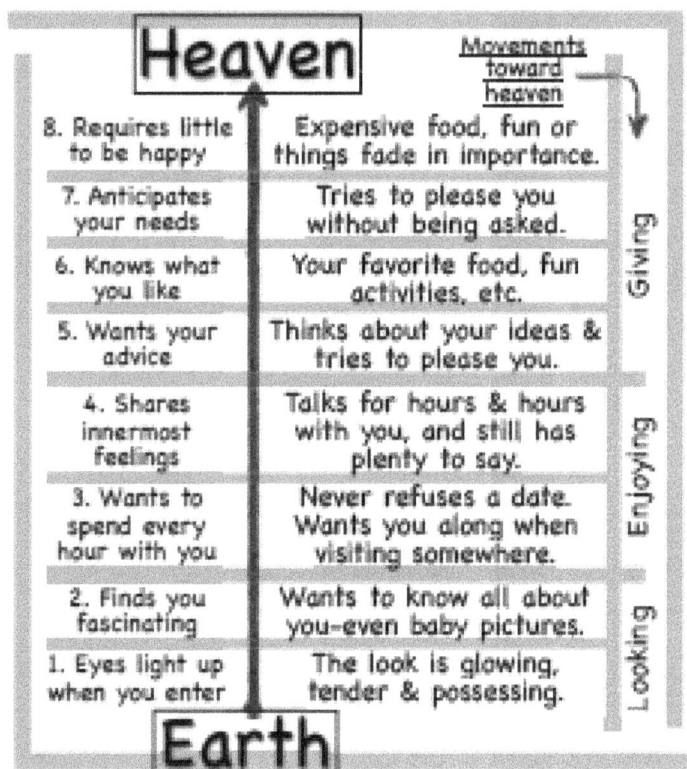

Heaven		Movements toward heaven
8. Requires little to be happy	Expensive food, fun or things fade in importance.	Giving
7. Anticipates your needs	Tries to please you without being asked.	Giving
6. Knows what you like	Your favorite food, fun activities, etc.	Giving
5. Wants your advice	Thinks about your ideas & tries to please you.	Giving
4. Shares innermost feelings	Talks for hours & hours with you, and still has plenty to say.	Enjoying
3. Wants to spend every hour with you	Never refuses a date. Wants you along when visiting somewhere.	Enjoying
2. Finds you fascinating	Wants to know all about you—even baby pictures.	Looking
1. Eyes light up when you enter	The look is glowing, tender & possessing.	Looking
Earth		

Love

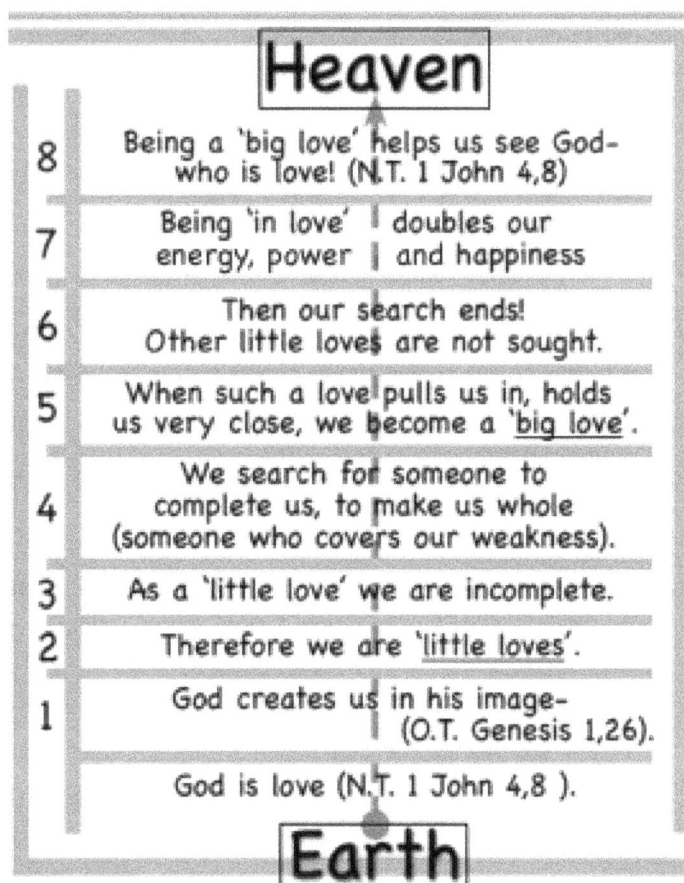

Heaven

8 — Being a 'big love' helps us see God- who is love! (N.T. 1 John 4,8)

7 — Being 'in love' doubles our energy, power and happiness

6 — Then our search ends! Other little loves are not sought.

5 — When such a love pulls us in, holds us very close, we become a 'big love'.

4 — We search for someone to complete us, to make us whole (someone who covers our weakness).

3 — As a 'little love' we are incomplete.

2 — Therefore we are 'little loves'.

1 — God creates us in his image- (O.T. Genesis 1,26).

God is love (N.T. 1 John 4,8).

Earth

Relationship

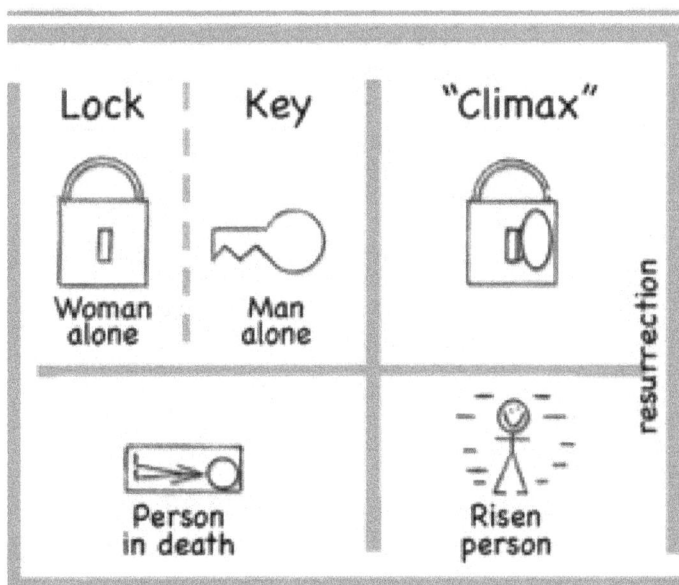

Eating Out

Summer ⟶	Fall ⟶	Winter ⟶	Spring
full life	fall in life	death	new life
Eating at home.	Tired of the same home meals.	Bored with home meals and kitchen work.	Eating out at a great, new restaurant.
Young person	Older person, tired	Person at death Lk 14,16 Rev 7,6	Risen person, enjoying heaven

Resurrection

Quarrels/Forgiving

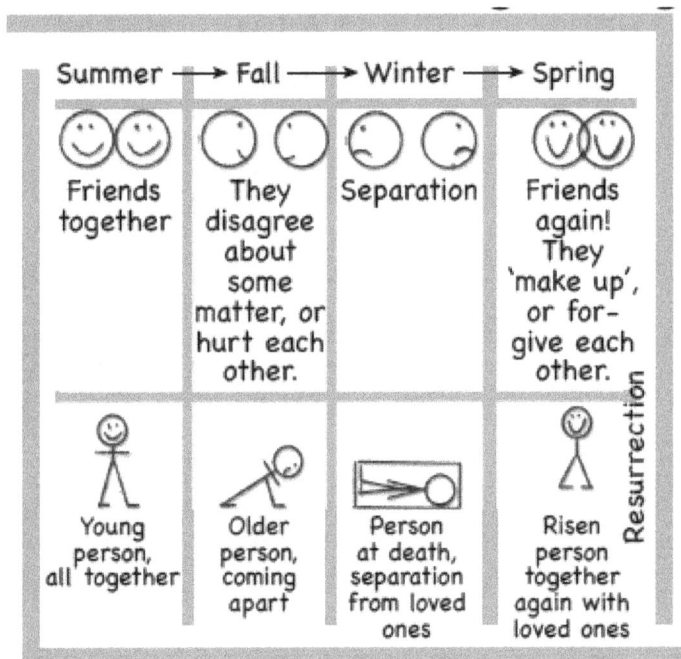

Summer →	Fall →	Winter →	Spring
Friends together	They disagree about some matter, or hurt each other.	Separation	Friends again! They 'make up', or for- give each other.
Young person, all together	Older person, coming apart	Person at death, separation from loved ones	Risen person together again with loved ones

Resurrection

Outcast

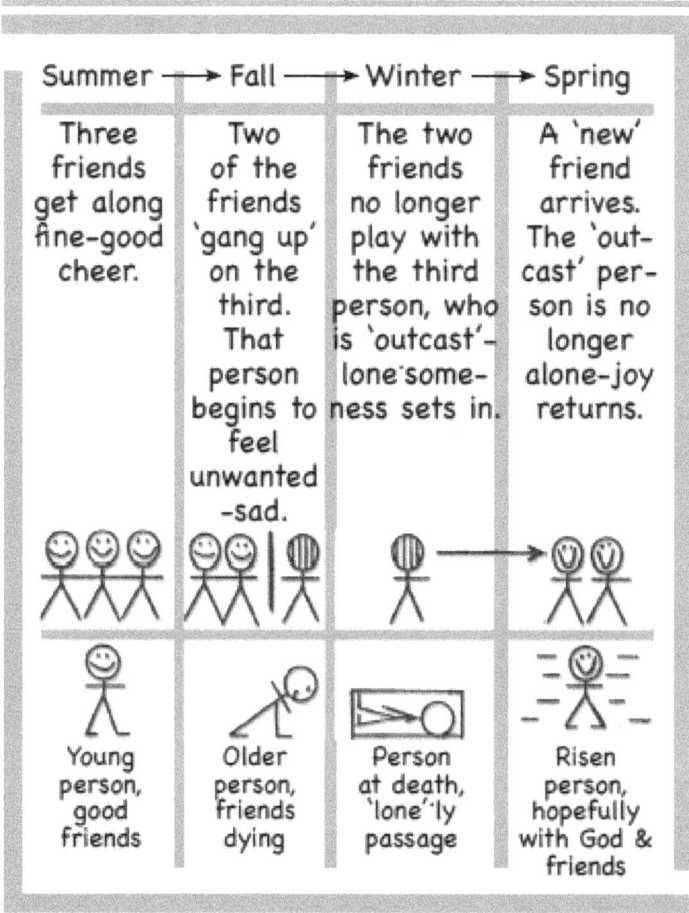

Summer →	Fall →	Winter →	Spring
Three friends get along fine-good cheer.	Two of the friends 'gang up' on the third. That person begins to feel unwanted -sad.	The two friends no longer play with the third person, who is 'outcast'- lone some- ness sets in.	A 'new' friend arrives. The 'out- cast' per- son is no longer alone-joy returns.
Young person, good friends	Older person, friends dying	Person at death, 'lone'ly passage	Risen person, hopefully with God & friends

The Poor

Helping the poor is a sign of resurrection. The poor can be raised up and given new life.

Mt 25, 35-40

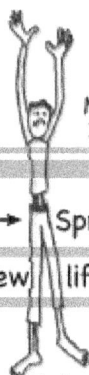

Summer →	Fall →	Winter →	Spring
full life	fall in life	death	new life
Person has a job, and good health.	Person has no job (or poor health). Result-no income.	Money runs out.	Help arrives! Family is raised up.
Family is well fed.	Hunger begins in family.	Starvation ensues, living death.	Food supply is restored.
Young person	Older person	Person at death	Risen person

Chapter 7

Signs of Risen Life
in the Human Family

Living Systems

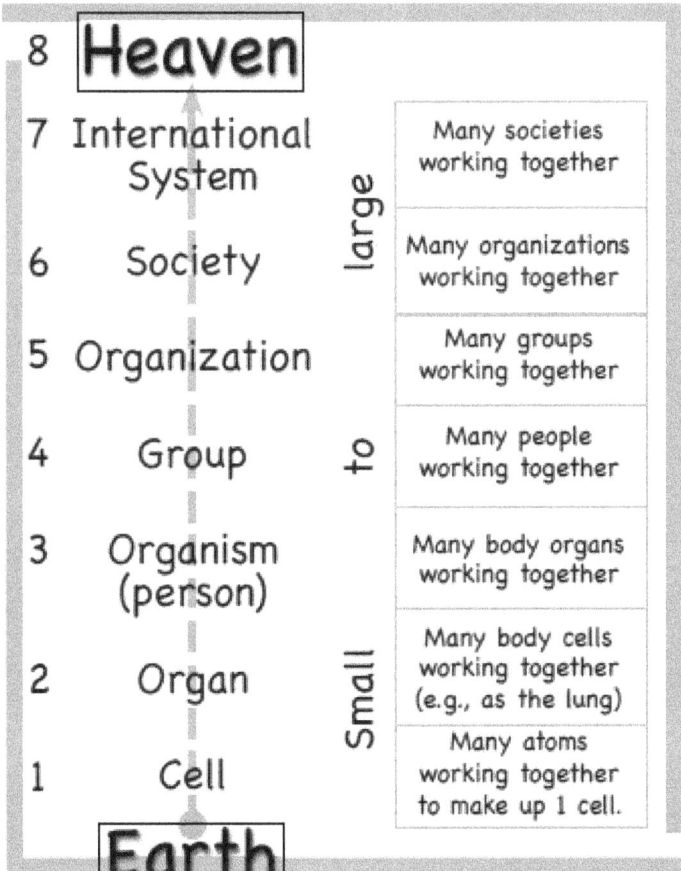

		large	
8	**Heaven**		
7	International System		Many societies working together
6	Society		Many organizations working together
5	Organization		Many groups working together
4	Group	to	Many people working together
3	Organism (person)		Many body organs working together
2	Organ		Many body cells working together (e.g., as the lung)
1	Cell	Small	Many atoms working together to make up 1 cell.
	Earth		

Marriage

Summer →	Fall →	Winter →	Spring
age 0-10	age 10-20	age 20	age 25
Child grows up within the family.	Teen'ager slowly cuts ties with his/her family.	Young adult leaves home. Can be a lonely, death-like time.	Young adult marries. A new family rises up.
Young person grows up within the world	Older person, slowly cuts off from world	Person at death, leaving this world	Risen person, beginning a new life, in heaven

Houses

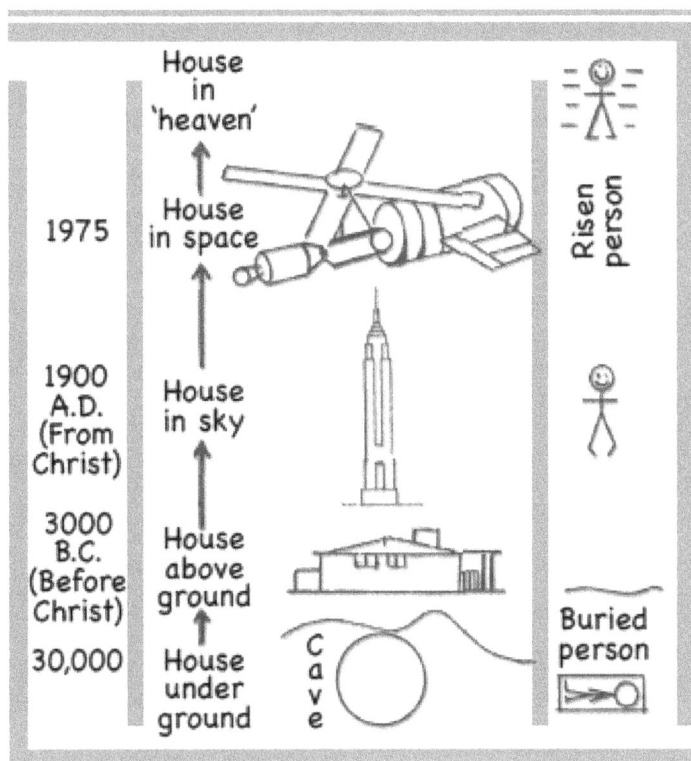

1975	House in 'heaven'		Risen person
	House in space		
1900 A.D. (From Christ)	House in sky		
3000 B.C. (Before Christ)	House above ground		
30,000	House under ground	Cave	Buried person

House Repairs

When things break down it is a sign
of death. When things are repaired
it is a sign of resurrection.

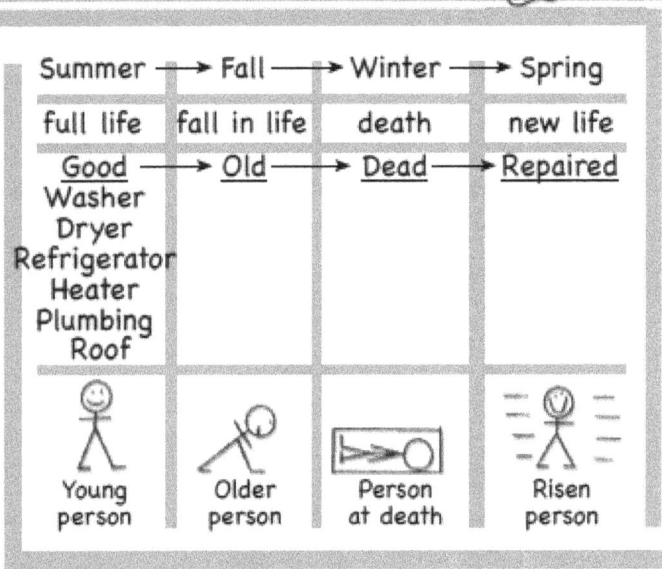

Summer →	Fall →	Winter →	Spring
full life	fall in life	death	new life
Good →	Old →	Dead →	Repaired
Washer Dryer Refrigerator Heater Plumbing Roof			
Young person	Older person	Person at death	Risen person

Changing Jobs

Summer →	Fall →	Winter →	Spring
full life	fall in life	death	new life
Job is going fine.	Employer warns of a possible 'layoff'.	Job is lost. Person is 'fired'.	A 'new job' is found.
Young person, doing fine.	Older person, 'laid up'.	Person at death. Life appears lost.	Risen person, with a 'new life'.

Moving

Summer →	Fall →	Winter →	Spring
Living in one house a long time. Many good friends are present.	Parents plan to move.	Family moves. Old friends are seldom seen again.	Family sets up a new house. New friends appear.
Young person	Older person, thinking about leaving earth.	Person at death, moving on. 2 Cor 5,1	Risen person, at home in heaven.

Resurrection

Joy

Summer →	Fall →	Winter →	Spring
Life is OK	Serious problem arises	Real loss occurs	Loss is overcome
	Possible loss of job, health, or house.	Fired from job, serious illness, house burns.	New job is found, health returns, house is rebuilt.
		Grief sets in.	Joy sweeps away grief.
Young person	Older person	Person at death	Risen person

Gatherings

Parting is a sign of death.
Coming together again is
a sign of resurrection.

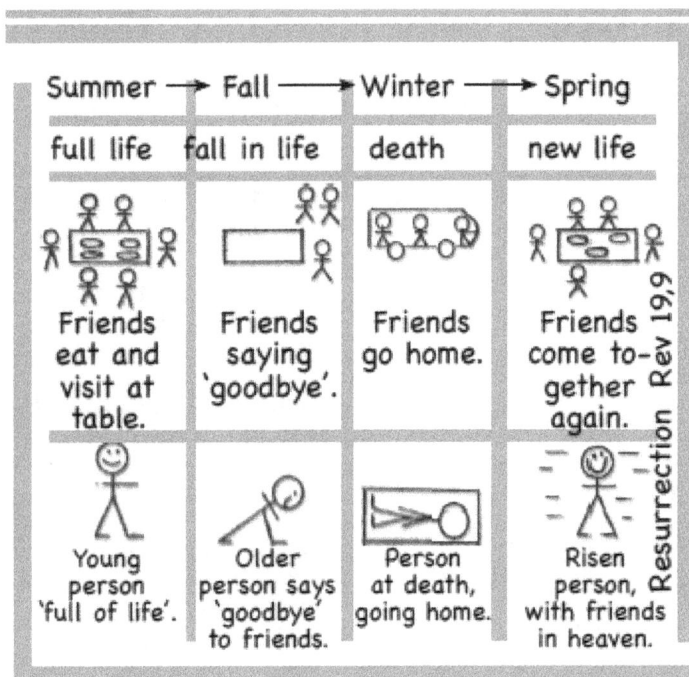

Summer →	Fall →	Winter →	Spring
full life	fall in life	death	new life
Friends eat and visit at table.	Friends saying 'goodbye'.	Friends go home.	Friends come together again.
Young person 'full of life'.	Older person says 'goodbye' to friends.	Person at death, going home.	Risen person, with friends in heaven.

Resurrection Rev 19,9

Holidays

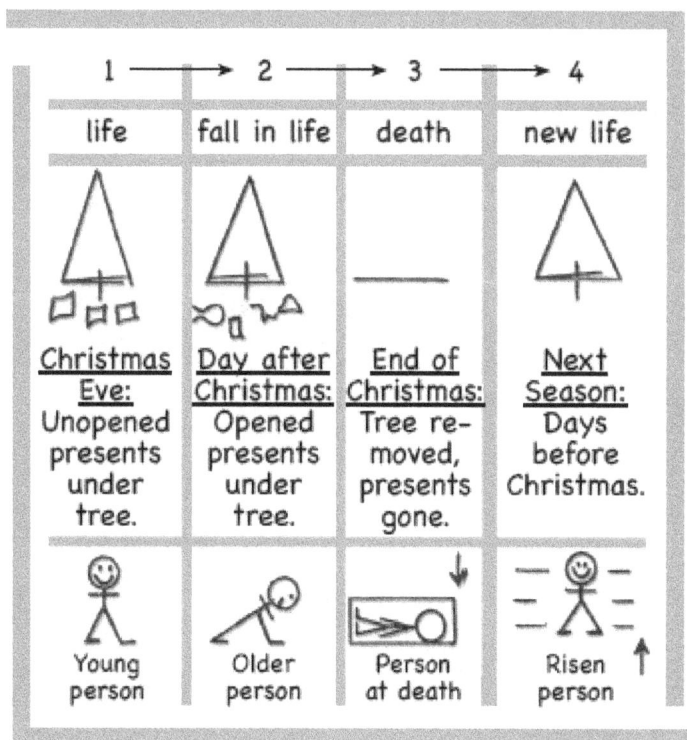

1 →	2 →	3 →	4
life	fall in life	death	new life
Christmas Eve: Unopened presents under tree.	Day after Christmas: Opened presents under tree.	End of Christmas: Tree removed, presents gone.	Next Season: Days before Christmas.
Young person	Older person	Person at death	Risen person

Music 1

Summer →	Fall →	Winter →	Spring
full life	fall in life	death	new life
Excited Life is good.	Bored Tired of the same old thing.	Depressed 'Sick to death' of ordinary days.	Uplifted 'Raised up' by great music! Music that uplifts the spirit.
 Young person	 Older person	 Person at death	 Risen person

Resurrection

Music 2

Great music is a sign of heaven: "as he drew near to the (Father's) house he heard music and dancing." Lk 15, 25

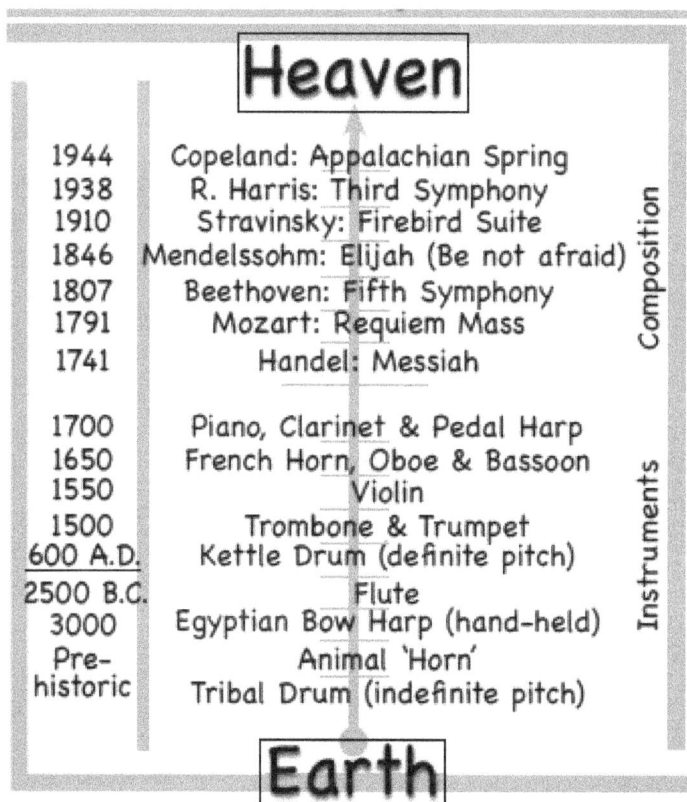

Heaven

1944	Copeland: Appalachian Spring	Composition
1938	R. Harris: Third Symphony	
1910	Stravinsky: Firebird Suite	
1846	Mendelssohm: Elijah (Be not afraid)	
1807	Beethoven: Fifth Symphony	
1791	Mozart: Requiem Mass	
1741	Handel: Messiah	
1700	Piano, Clarinet & Pedal Harp	Instruments
1650	French Horn, Oboe & Bassoon	
1550	Violin	
1500	Trombone & Trumpet	
600 A.D.	Kettle Drum (definite pitch)	
2500 B.C.	Flute	
3000	Egyptian Bow Harp (hand-held)	
Pre-historic	Animal 'Horn'	
	Tribal Drum (indefinite pitch)	

Earth

Worship

To worship is to hear and praise God
in community with other people.

Death ──────▶ Resurrection	
Without worship life is less meaning'ful, dark.	Worship lifts the spirit! A little of heaven is experienced.

Vacations

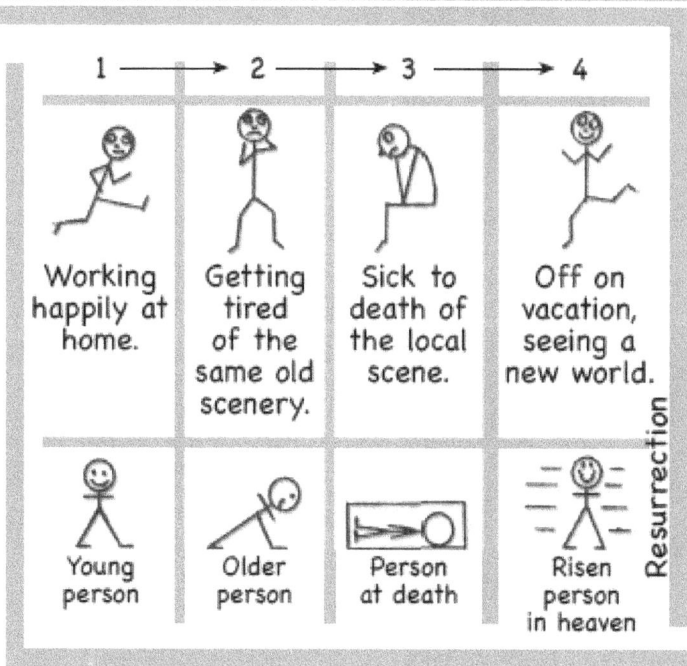

1 →	2 →	3 →	4
Working happily at home.	Getting tired of the same old scenery.	Sick to death of the local scene.	Off on vacation, seeing a new world.
Young person	Older person	Person at death	Risen person in heaven

Resurrection

Lasting Joy

On earth there are all sorts of things that promise to give us joy. But they never quite keep their promise! If I find in myself a desire which no experience can satisfy, the most likely explanation is that I was made for another world.

	Heaven	Lasts forever because God is its source
(◡)	Perfect joy	
(◡)	Great joy	Long term in memory
(◡)	more joy	Short term in memory
(◡)	Some joy	Short lived (forgotten almost immediately)
	Earth	

Ongoing Life

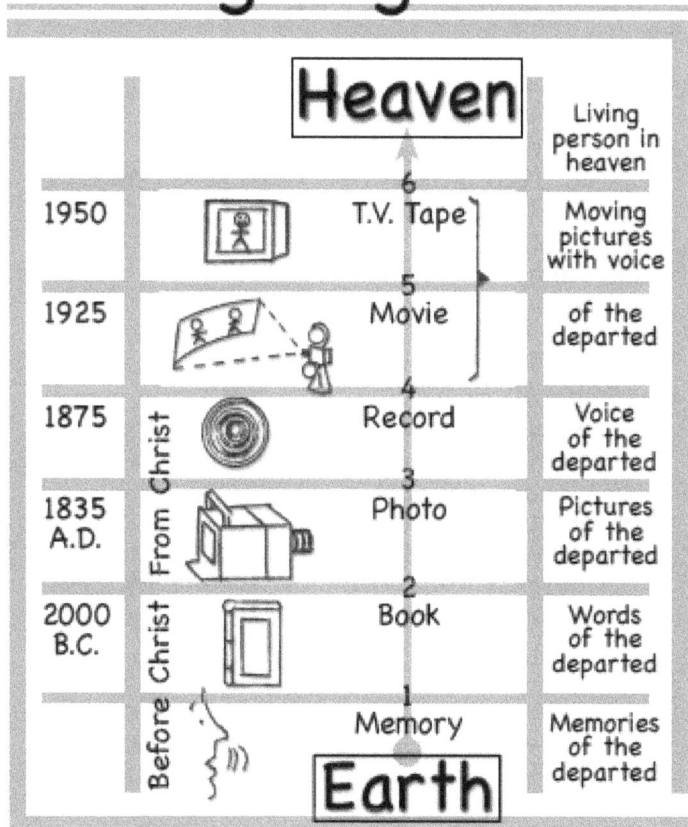

		Heaven		Living person in heaven
			6	
1950		T.V. Tape		Moving pictures with voice
			5	
1925		Movie		of the departed
			4	
1875		Record		Voice of the departed
			3	
1835 A.D.		Photo		Pictures of the departed
			2	
2000 B.C.		Book		Words of the departed
			1	
		Memory		Memories of the departed
		Earth		

From Christ

Before Christ

Short-Term Death

Death ⟶ Resurrection	
People are de-clared dead in hos-pitals. They have no heartbeat, breathing, or brain waves. Doctors work to revive them. Then after 5, 10, 50 or 100 minutes they are brought back to life. The question is: What happened to them when they were dead? (No. 1-5 gives the main points)	1. Person finds himself outside his physical body looking down upon his doctors.
	2. Very soon dead relatives greet the man.
	3. Then God comes as the "Being of light".
	4. Next the dead man is shown a flash-back of his life.
	5. Lastly he is called back into his body: before reward or punishment is given.
Person at death	Risen person 2 Cor 5,8

Risen Body

Summer →	Fall →	Winter →	Spring
full life	fall in life	death	new life
Physical body Spirit	1 Cor 15,44	2 Cor 5,1	Lk 20,36 & 24,31
Our house is like our physical body.	Living in the house is like our spirit living in our body.	When we move out of the house (or body) it begins to fall apart.	Our new house (or spiritual body) never dies in heaven.
Young person	Older person	Person at death	Risen person

Chapter 8

Signs of Risen Life
Christ

God's Care

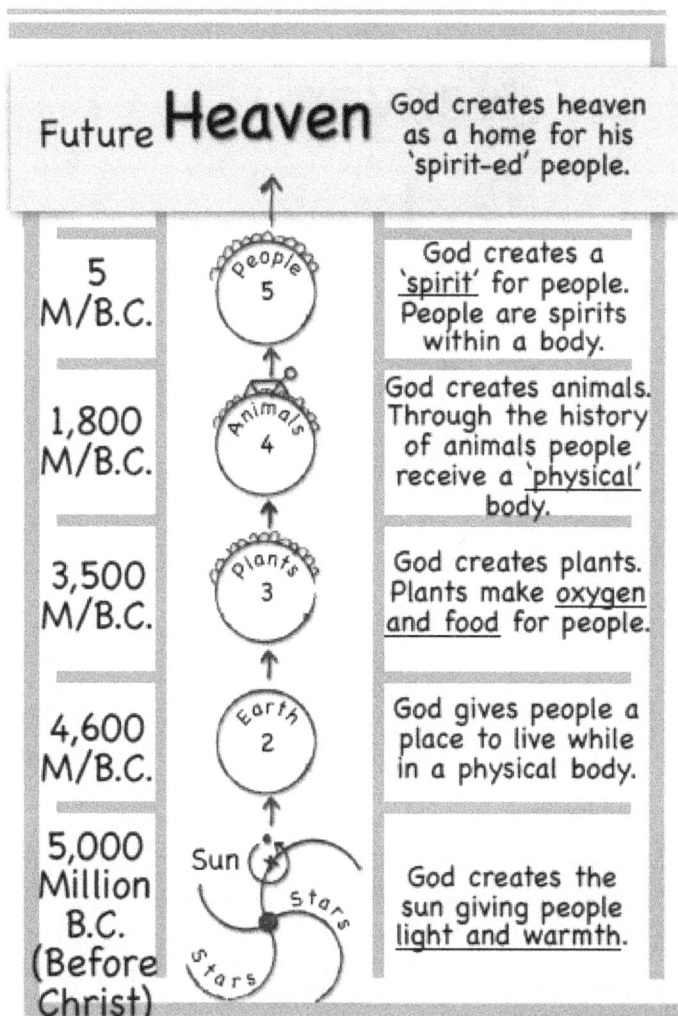

Future Heaven God creates heaven as a home for his 'spirit-ed' people.

5 M/B.C.	People 5	God creates a 'spirit' for people. People are spirits within a body.
1,800 M/B.C.	Animals 4	God creates animals. Through the history of animals people receive a 'physical' body.
3,500 M/B.C.	Plants 3	God creates plants. Plants make oxygen and food for people.
4,600 M/B.C.	Earth 2	God gives people a place to live while in a physical body.
5,000 Million B.C. (Before Christ)	Sun Stars Stars	God creates the sun giving people light and warmth.

Direction

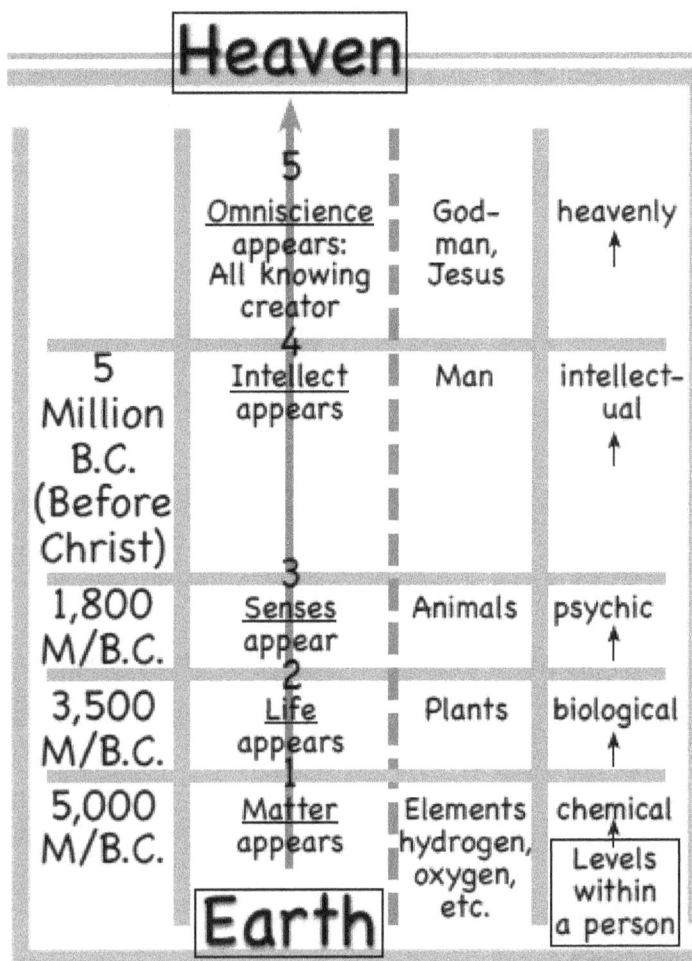

	Heaven		
	5 Omniscience appears: All knowing creator	God- man, Jesus	heavenly ↑
5 Million B.C. (Before Christ)	**4** Intellect appears	Man	intellect- ual ↑
1,800 M/B.C.	**3** Senses appear	Animals	psychic ↑
3,500 M/B.C.	**2** Life appears	Plants	biological ↑
5,000 M/B.C.	**1** Matter appears	Elements hydrogen, oxygen, etc.	chemical ↑ Levels within a person
	Earth		

Bible

Heaven

God's Kingdom (N.T.)
Lk 24,27

Com-pletion	Good servants watch for God's coming.
Duties	Good Samaritans help needy. Good trees produce good fruit.
Mem-bership	It is like a marriage feast, or great supper.
Its value	The heavenly kingdom is a priceless pearl (Jesus-Mt 13,46)

Human Kingdom (O.T.)

70 A.D. (From Christ)	The O.T. history is ended by Roman armies.
586 B.C.	Jeremiah warns! Southern kingdom is ended by the Babylonians.
721 B.C.	Amos & Isaiah warn! Northern kingdom is ended by the Assyrians.
921 B.C.	Kingdom is divided by civil war under Solomon's son.
1040 B.C.	Samuel Warns! But Saul-David set up a human kingdom.
1445 B.C.	Moses leads God's people into the Holy land.
1800 B.C. (Before Christ)	Abraham receives God's promise of an 'everlasting' people. *Acts 28,23*

Earth

Holy Days

Summer →	Fall →	Winter →	Spring
full life	fall in life	death	new life
June-Aug	Sept-Nov	Dec-Feb	Mar-May
1. 75°F	50°F	25°F	50°F
2. Daylight: 16 hours	Daylight: 12 hours	Daylight: 8 hours	Daylight: 12 hours
3. Soft ground	Hardening ground	Hard ground (buried under snow)	Softening ground
4. Many leaves	Dying leaves	Leaves gone	New leaves
5. Many birds	Fewer birds	Few birds	Birds return
6. Mammals are active	Mammals are less active	Mammals sleep under-ground	Mammals wake
The invisible spirit prepares the earth for God's visible coming		Christmas → In the season of death Jesus comes-to give life	Easter Jesus rises from the dead, showing us the way

Jesus' Power

Our body is 2/3 water.
Jesus showed power over water,
by walking upon the Sea of Galilee.
He promised to use this power
on our body when we die.

Mt14,23

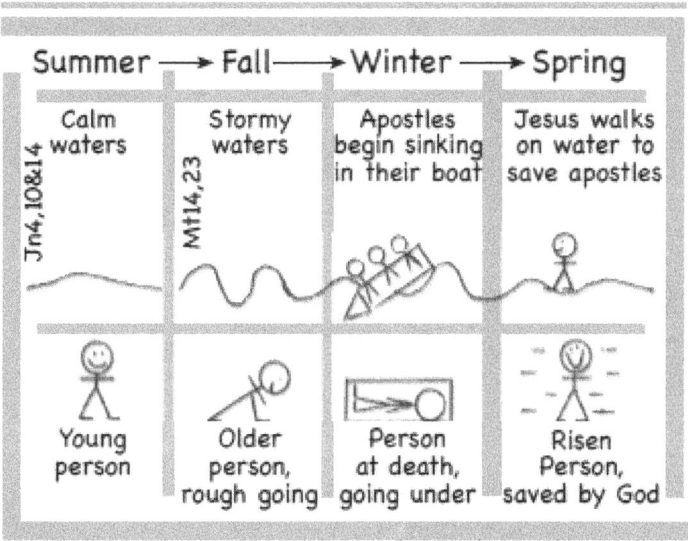

Summer →	Fall →	Winter →	Spring
Calm waters	Stormy waters	Apostles begin sinking in their boat	Jesus walks on water to save apostles
Jn4,10&14	Mt14,23		
Young person	Older person, rough going	Person at death, going under	Risen Person, saved by God

Jesus Raises Dead

Widow's son

Luke 7,11

Jesus has power over death.

Jesus restored natural life

to the widow's son.

He promises risen life to us.

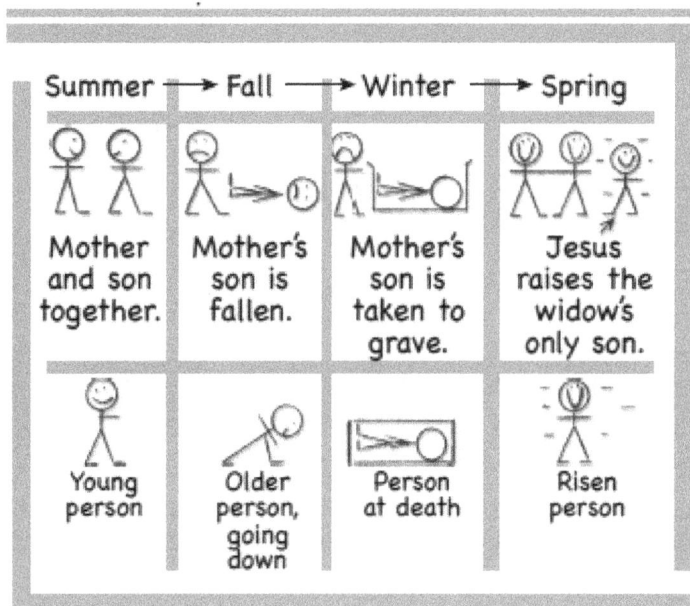

Summer →	Fall →	Winter →	Spring →
Mother and son together.	Mother's son is fallen.	Mother's son is taken to grave.	Jesus raises the widow's only son.
Young person	Older person, going down	Person at death	Risen person

Jesus' Resurrection

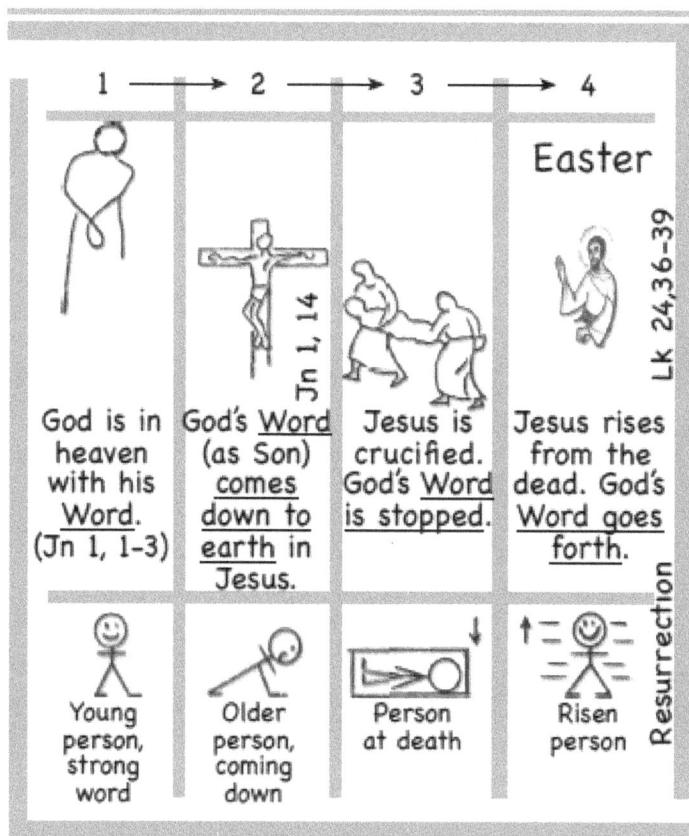

1 ──▶ 2 ──▶ 3 ──▶ 4			
			Easter
God is in heaven with his Word. (Jn 1, 1-3)	God's Word (as Son) comes down to earth in Jesus.	Jesus is crucified. God's Word is stopped.	Jesus rises from the dead. God's Word goes forth.
Young person, strong word	Older person, coming down	Person at death	Risen person

Jn 1, 14

Lk 24,36-39

Resurrection

Final Thoughts

The resurrection of Jesus was a great upwelling of love for humankind. It blanketed the world with a new kind of certainty, revealing the original intentions of God. Jesus rose from the dead in a real but empowered human body, showing humans who have been created in his own image, a way of life beyond death (Genesis 1:27).

When Jesus rose from the dead, he exited his tomb through its walls. The stone enclosure in front of the opening of the tomb was later removed by an angel so that certain women could enter and anoint his body.

> "an angel of the Lord descending out of heaven … rolled away the stone and sat upon it."
> - Matthew 28:2

Christ himself not only rose from the dead on Easter Sunday morning, but he remained with his Apostles and early followers as a living example for some forty days after his resurrection.

"He presented himself alive … through <u>forty days</u>, being seen by them and speaking of the things concerning the Kingdom of God."
- Acts 1:3

"to witnesses … to us who ate and drank with him <u>after he had risen</u> out of the dead …"
- Acts 10:41

After rising on Easter Sunday morning, Jesus appeared to two of his disciples about noon on that same day. He walked with them from Jerusalem to Emmaus, a distance of about seven miles. These disciples <u>did not recognize Jesus</u> in his risen body until later that afternoon. As evening arrived, at supper time, Jesus took bread and said the transforming, sacred words, first heard at the Last Supper (Luke 22:19).

"he [Jesus] took bread [at the Last Supper] blessed it and broke it, and gave it to them, saying: 'This is [now] my Body, which is given for you. Do this action [shown here] in remembrance of me."
- Luke 22:19

That evening, after the bread had become the Eucharist (the words of consecration having been said), the Emmaus disciples finally recognized who Jesus was. Then, at that moment, and only at that moment, did Jesus disappear from their sight!

> "Their eyes were opened up [upon being in the presence of the Eucharist], and they recognized him [in his risen body]; and [then] he vanished out of their sight."
> - Luke 24:31

In other words, Jesus was present to them in another bodily form (the Eucharist). He no longer needed to be present to them in his risen, physical body! So, he physically disappeared!

Later, on that same evening of Easter, the risen Jesus also came through closed doors to his eleven Apostles back in Jerusalem. They were at first startled by him, thinking him to be a ghost, but then, when they had examined his wounds and Jesus had eaten some fish, it became clear to them that his body was truly physical in nature and that he was really back from the dead!

Fast forward to forty days after Jesus's resurrection from the dead. At that time, Jesus left this planet in what is now called his "Ascension" into Heaven (John 20:17; Luke 24:51; Acts 1:9; Mark 16:19). In other words, Jesus did not die a second time! Jesus had promised his Ascension earlier in his ministry, after he had miraculously fed some 5,000 people. To certain doubters, at that time, he said:

> "What then if you see the Son of Man [myself] ascending to where he was at first?"
> -   John 6:62

Later still, some fifty days after his resurrection, Jesus sent the Holy Spirit to guide his Church on its future earthly journey to Heaven (Ephesians 1:23). On that day, called Pentecost, the Apostles baptized some 3,000 people after Peter's sermon (Acts 2:41). These people were not necessarily baptized because they had heard Peter's sermon! Rather, they were baptized because they had seen, or even possibly had talked with Jesus, during the forty days after his resurrection from the grave. Large numbers of people had long been following this "startling person" whom certain men in power had put to death. Large numbers of ordinary people had seen the resurrection, or knew of its reality. That explains why the temple officials tried to hide the fact of Jesus' resurrection from an even wider circle of interested persons (Matthew 28: 11-15).

A short while after Pentecost, Peter and John were again "teaching the people, announcing the resurrection of Jesus from the dead" ... and as a result, another 5,000 believed – and were baptized (Acts 4:2-4). The Apostles had no microphones or sound system to reach such large numbers of people. These new believers, like the Apostles themselves, had seen the risen Jesus. That is why so many were entering the Church!

Twenty years later, in 53 AD, St Paul wrote his first letter to the Christians at Corinth in Greece. In this letter, Paul spoke of some of the early witnesses to Jesus's resurrection.

"he [Jesus] was buried and rose on the third day according to the Scriptures [of Matthew and Luke], and he was seen by Cephas [St. Peter], then by the

Twelve [Apostles]; afterward he was seen by
over 500 brothers at one time, of whom the
majority remain until now, though some have
fallen asleep [died]. Afterward he was seen by
James ... and last of all he was also seen by me
[Acts 9:3-5; 22:6-10]."
- 1 Corinthians 15:4-8

Finally, in 107 AD, a mere 75 years after the
resurrection, St. Ignatius of Antioch (35-107 AD)
spoke of the Church as having spread throughout
the Roman world. Ignatius, who was himself a
bishop, died a martyr in the Colosseum in Rome. He
wrote seven letters while being transported to
Rome. In his letter to the Christians at Ephesus, he
spoke of the widespread presence of the Church by
107 AD.

"bishops who are stationed everywhere into the
farthest parts [of the Roman Empire] are [doing]
the will of Jesus Christ."
- Letter to Ephesians 3

Thus, one can see that Jesus's words, miracles and resurrection had spread quickly across the Roman world.

The objective of this book has been to show that the model of "resurrection" is present in all of nature. Nature is a fundamental witness to the work and person of Jesus. He is the author of the world of nature, and as such he has left clues in the fabric of created things, that point to his own resurrection and power over death.

"We understand that the ages were put in order by the word of God, so that the things being seen have come out of things not appearing."
- Hebrews 11:3

"He [Jesus] is the image of the unseen God [the Father] ... all things have been created through him [as God's Word] ... He himself is before [the creation of] all things, and in him all things [presently] hold together."
- Colossians 1:15-17

In conclusion, one should not be surprised by the many "signs of resurrection" on this planet, which together point to God's protective intention for each one of us!

www.ingramcontent.com/pod-product-compliance
Lightning Source LLC
Chambersburg PA
CBHW070013110426
42741CB00034B/1623